T0189949

Ready

Roberto Dillon

Ready

A Commodore 64 Retrospective

 Springer

Roberto Dillon
James Cook University
Singapore
Singapore

ISBN 978-981-10-1257-0   ISBN 978-981-287-341-5   (eBook)
DOI 10.1007/978-981-287-341-5

Springer Singapore Heidelberg New York Dordrecht London
© Springer Science+Business Media Singapore 2015
Softcover reprint of the hardcover 1st edition 2015

Printed on acid-free paper

Springer Science+Business Media Singapore Pte Ltd. is part of Springer Science+Business Media
(www.springer.com)

To all my school friends with whom I shared the early passion for 8-bit computers, in particular:
Alberto Baratta, Francesco Gambaro,
Andrea Lombardo, Matteo Lunardi,
Silvio Morando, Mario Oneto, Marco Sardi
and in memoriam of Lorenzo Borghetto
(1973–2002)

Foreword

How Commodore Opened the Floodgates to Home Computing

Today, as most of us use smartphones and other devices that are genius-levels smarter than the first home computers, we tend to forget where home computing got started.

In this excellent, well-researched and entertainingly written book, Roberto Dillon tells the story of how the Commodore 64 (and its predecessor, the Commodore VIC-20) jump-started the home computer revolution. You will learn how home computing got its start, and gain new perspectives on the home computer revolution that opened the floodgates to the remarkable computing technologies and apps we are benefiting from today.

Roberto's experience receiving his first home computer—a Commodore 64—at the age of 10 is a story that was repeated all over the world in the 1980s. Tens of millions of people, mostly in the U.S. and Europe, were introduced to home computing by Commodore, thanks to company founder Jack Tramiel and a brilliant team of innovative engineers and marketers who unleashed the home computer revolution. I was privileged to play a role in the design and launch of those first home computers and I have to say that I am impressed by the scope and insights presented in this terrific book by Professor Dillon.

Roberto describes how the Commodore 64 jump-started the home computer revolution. From the unique and inventive architecture of the computer itself, to creative marketing (using William Shatner as Commodore's spokesman), to the thousands of applications developed by an enthusiastic user community, Roberto gives Commodore its rightful place in the history of home computing. He draws a direct line from the C64 to many of the technologies and apps we take for granted today, from video games and telecomputing to early applications that spawned the Internet.

He also gives much-deserved credit to the engineers and marketers who developed this remarkable pioneering computer, and to Jack Tramiel, the visionary who wanted to design computers "for the masses, not the classes."

I was fortunate to play a role in the development of the Commodore VIC-20 and Commodore 64, and I can attest to the electric, innovative environment that existed in the company. We were conscious that what we were doing could change the world. Today, I still receive emails from people around the world expressing their gratitude and describing how their first Commodore computer transformed their lives.

In the early 1980s, everyone was asking, "Where's the home computer revolution? When does it start?" Businesses used minicomputers and mainframes. Most schools could not afford computers for their students. Then Commodore computers came along with sleek, Porsche-inspired designs, affordable prices, and the power needed to drive practical applications like wordprocessing and spreadsheets, that we all take for granted now.

When we were developing the first home computers, there were no cellphones or wireless networks because the wireless infrastructure did not exist. Commodore's plug-in modem, which I co-designed, and the Commodore Information Network which I established on CompuServe, were among the first telecomputing networks, an early precursor to Internet communities.

Making home computers affordable was a huge achievement. Suddenly, for a few hundred dollars, anyone could use powerful apps that were previously available on minicomputers and mainframes only. Hobbyists could write their own software programs. Commodore computers quickly found their way into classrooms—even kindergartens—where the first generation of students were introduced to computing.

These achievements occurred more than 30 years ago but they are not forgotten. As Roberto observes, Commodore today is not an artefact, but a living, dynamic technological heirloom that has found a home among retrocomputing enthusiasts and collectors of computer memorabilia. The Commodore 64 is on display in many museums around the world while dozens of retrocomputing clubs keep alive the "Commodore spirit" by restoring and maintaining original computers, celebrating the achievements by the talented and dedicated people who launched the home computer revolution.

It is extremely gratifying to see that the Commodore spirit is alive and well, and new "Commodorians" are continuing the tradition of innovation that we began, so many years ago.

Michael Tomczyk
Home Computer Pioneer
Innovator in Residence, Villanova University

Acknowledgments

I am highly indebted to Mr. Michael Tomczyk and Mr. Nigel Parker for their time and fundamental help. I am also grateful to the editorial staff at Springer, in particular Mr. Loyola D'Silva and Ms. Amudha Vijayarangan, for believing in this project and making it a reality.

Last, but clearly not least, special thanks also to all the C64 hobbysts, programmers and hackers around the world who managed to keep our shared passion alive for all these years, with a special mention to the Lemon64 community for the inspiring and insightful discussions on its forum!

Contents

About the Author

Roberto Dillon originally from Genoa, Italy, but currently based in Singapore, Dr. Roberto Dillon is active both as a developer and as a professor of game design.

As an academic and educator, Roberto was the Game Design Department Chair at DigiPen Institute of Technology Singapore, with several of his students gaining top honors at the Independent Game Festival (IGF) both in San Francisco and Shanghai, and is now an Associate Professor lecturing Game Design in James Cook University, where he is also the Curator of the 'JCU Museum of Video and Computer Games', the first museum completely dedicated to the preservation of video games in South-East Asia. As a developer, on the other hand, his games have been showcased at events like Sense of Wonder Night in Tokyo, FILE Games in Rio de Janeiro and at the Indie Prize Showcase Asia, besides reaching top positions on Apple's AppStore across several countries.

Roberto is a regular speaker at game related conferences worldwide and wrote different books for AKPeters and CRC Press: 'On the Way to Fun', where he introduced the 6–11 Framework, a game design methodology now referenced in several university curricula as well as used by game designers in small and big studios alike, 'The Golden Age of Videogames', a history of the gaming industry from its origins till the PlayStation launch and 'HTML5 Game Development from the Ground Up with Construct 2' to introduce aspiring developers to the world of game making.

"Ready: A Commodore 64 Retrospective", published by Springer, is his latest work on the history of home computing and its impact on modern society.

Introduction

It was a rainy day in late November 1981 when I, a young eight-year-old boy, followed my parents who were invited for dinner by one of my dad's colleagues and mentors, Prof. Augusto Gamba (1923–1996), for an evening that, unknowing to all of us, was going to set the future direction of my life.

Professor Gamba, a distinguished theoretical physicist author not only of influential scientific works but also of brilliant educational books, had just got a new computer from the US, a TI-99/4A, and was willing to show us this little new technological marvel. Needless to say, I have no memory at all of what was on the dining table that evening: all my attention, in fact, was quickly grabbed by the home computer and the programs Prof. Gamba showed us. On the way back home I was holding tight in my tiny hands a roll of paper with abstract geometrical patterns printed on the TI's plotter: I remember staring at them in wonder for a long time and I couldn't really sleep well that night. I was too excited: in those drawings, in those sounds and images coming from the computer monitor, I actually saw my own future and knew that computers were going to play a fundamental role in my life.

As you can imagine, it did not take long before I started bugging my dad to bring a computer home. Finally, for my tenth birthday in 1983, we went to a nearby electronics shop and got back home with a computer that had just been released in Italy: the Commodore 64. This was the beginning of my lifelong learning journey with technology and programming: I started learning BASIC to develop my own games, first with the help of my dad and then on my own. These were simple arcade style games or text adventures with rudimentary graphics that only my closest friends ever played or pretended to care about. Nonetheless, the die was cast and the rest, so to say, is history.

Those pioneering years were a truly magical and revolutionary time. I have no doubt that the "Computer Revolution" that happened across the decade spanning the late 1970s till the late 1980s, with computers getting into the homes of millions worldwide, will be regarded in the history of mankind with the same importance, if not even more, as the Industrial Revolution of the nineteenth century. It was something that changed the world and I consider myself very lucky for having lived at such a time.

Among all the different home computers, the Commodore 64 holds a special place in the hearts of many, being the best selling model of all time. This book aims at discussing both its origins and the characteristics that made it so special and successful. Most importantly, it aims at discussing and analysing the different ways it affected the rise and popularity of home computers, together with the related game industry. In fact, not only innovative, genre defining games were born or became popular on this platform but also user friendly development tools, like the first generation of what was going to be later called "game engines", were made widely available to the general public. These programs allowed a generation of passionate kids to easily experiment with their ideas, learning the ropes of game design and possibly seek fame and fortune, paving the way for modern "indie" (independent) game developers and corresponding gaming culture.

Nonetheless, focusing exclusively on games would be reductive as the C64 influence on modern technology runs much deeper: affordable GUI-based Operating Systems and new modems, bringing many people online for the first time, were also fundamental to the growth and acceptance of computers as must-have home appliances. These are discussed here as well to better appreciate the overall impact Commodore in general and the C64 in particular had.

Overall, whether you are a C64 original user like myself or a much younger enthusiast who wants to learn about the origins of our modern computer-driven society, I am confident you will find something interesting and new here. Either way, if reading this book will bring back good old memories or spawn the curiosity for digging deeper into the subject, my mission would be accomplished and the long hours passed in writing this book would have been worth every minute.

Singapore, August 2014   Roberto Dillon

Chapter 1
Computers for the Masses, not the Classes

Abstract This chapter discusses the origins of Commodore under the helm of Jack Tramiel and the events that followed MOS acquisition, from the release of the PET to the idea of providing "computers for the masses, not the classes" which led to the VIC-20 and, ultimately, the C64.

Keywords Jack Tramiel · Commodore · MOS · 6502 · KIM-1 · PET · VIC-20

A lot has been written about Jack Tramiel (1928–2012) and, undoubtedly, much more is yet to be written about a self made business man with an iron fist who managed to build an empire and ignite a technological revolution out of nothing.

Born in Lodz, Poland, as Idek Tramielski (or Jacek Trzmiel, according to different sources) he emigrated to the USA after surviving the horrors of Nazi's concentration camps in Auschwitz during World War II. In 1952, after learning how to repair office equipment while serving in the US Army, he started out his own small typewriter repair shop in the Bronx, New York, while also working as a taxi driver. Then, in 1954 he founded the "Commodore Portable Typewriter Company" and moved to Toronto, Canada, to assemble and sell typewriters imported from Eastern Europe, something that wasn't legally possible in the US during those cold-war years. The company was then renamed to "Commodore Business Machines" (CBM) when formally incorporated in 1955.

However, in 1966 the pressure from cheaply imported Japanese typewriters, together with a financial scandal, almost put CBM out of business and only the intervention of a Canadian investor named Irving Gould (1919–2004) managed to keep the company afloat. Gould bought 17 % of the shares, becoming Commodore's Chairman. Under his influence the company shifted focus from typewriters to calculators, moving back to the USA in 1968 with a new headquarter in California.

Unfortunately, savage competition and price wars led once again by Japanese calculator manufacturers as well as Texas Instruments, brought Commodore on its knees in the mid 70s.

It was at this critical time, in 1976, that Jack made one of the most significant business decisions of the century, asking Gould to finance the acquisition of MOS

Technology with a $3 million investment that enabled Commodore to implement its own "vertical integration" strategy, i.e. build as many components as possible for its own products in-house, without relying on external vendors.

At the time, MOS was a relatively small semiconductor and chip manufacturer that, thanks to the genius of Chuck Peddle (1937) and a team (1937) and a team of extremely bright engineers who had previously worked together on the first 6800 Motorola CPU, had just released a new microprocessor: the 6502. The new CPU was designed to be as flexible as possible to suit applications for a multitude of devices and home appliances and was revolutionary in many respects, including its price: at a time when Motorola and Intel were selling their 6800 and 8080 CPUs respectively for a few hundred dollars apiece, the 6502 was available for $25 only! This astonishing feat was made possible also thanks to a unique manufacturing process in place at MOS that allowed for producing chips with a much higher success rate: in those pioneering days, in fact, up to 70 % of manufactured chips at Motorola and Intel were defective and had to be thrown away, rising the final costs for the good ones left.

Anyway, to capture people's attention and show the 6502 wasn't a toy, like some people suspected at first due to its ridiculously low price, Peddle and his team also went on to design a dedicated computer based on it, the KIM-1 (short for "Keyboard Input Monitor", see Fig. 1.1).

Although the KIM-1 was a very simple computer with only a 1 KB RAM, it managed to offer very handy features like saving programs to a tape device. This, together with an extensive documentation, made it relatively user friendly (for the time) and made the computer a perfect tinkering machine for the new, burgeoning group of computer hobbyists and hackers, enabling them to experiment freely and learn how to use the new tool.

MOS started selling the KIM-1 towards the end of 1975 and, despite its essential simplicity, it deserves a very important place in history since it was the world's first single board microcomputer: like later home computers, all main components were mounted on a single motherboard. This was something quite revolutionary at the time if we consider that all other contemporary microcomputers, like the Altair 8800 (Fig. 1.2), were designed instead as a set of interconnected boards, each hosting components dedicated to specific tasks, like memory, CPU, I/O interfaces etc.

Commodore and MOS were a perfect match and, under the influence of Peddle, Jack soon agreed Jack soon agreed to start diversifying Commodore's portfolio, progressively reducing its involvement in calculators and moving into fully fledged computers instead.

The first fruit of this new focus was the PET (Fig. 1.3), short for "Personal Electronic Transactor", just a made-up name to justify the cute acronym. The PET was the first all-in-one personal computer integrating keyboard, monitor and cassette recorder for storing and loading programs.

The PET was first publicly presented in January 1977 at the Winter Consumer Electronics Show (CES) and later at the West Coast Computer Faire, where also Steve Jobs and Steve Wozniak were presenting their seminal Apple II, running a

Fig. 1.1 Tens of thousands of KIM-1s were sold to budding engineers from its debut in late 1975 till 1980. Here one of the first ads published in computer magazines in 1976

MITS

BUILDING
YOUR OWN COMPUTER
WON'T BE A PIECE OF CAKE.

(But, we'll make it a rewarding experience.)

Chances are you won't be able to assemble the *Altair 8800 Computer* in an hour or two. But, that's only because the *Altair* is a real, full-blown computer. It's not a demonstration kit.

The *Altair Computer* is fast, powerful, and flexible. Its basic instruction cycle time is 2 microseconds. It can directly address 256 input and 256 output devices **and** up to 65,000 words of memory.

Thanks to buss orientation and wide selection of interface cards the *Altair 8800* requires almost no design changes to connect with most external devices. Up to 15 additional cards can be added inside the main case.

The *Altair Computer* kit is about as difficult to assemble as a desktop calculator. If you can handle a soldering iron and follow simple instructions, you can build a computer.

You see, at *MITS*, we want your experience with our kits to be rewarding. That's why we take such pains to write an accurate, straight-forward assembly manual. One that you follow step-by-step. (We leave nothing to the imagination.)

Some electronic kit companies are experts at cutting the corners. They promise you the sky and deliver a box full of surplus parts and a few pages of faded instructions run off on their copying machine.

We're experts at **not** cutting the corners. Our *Altair Computer* has been designed for both the hobby and the industrial market. It has to be constructed of the finest, quality parts. And it is.

That's why we give you double-sided boards, gold-plated connectors, a 10 Amp power supply (enough to power 15 additional cards), toggle switches and an all aluminum case complete with sub-panel and detachable dress panel.

That's why we give you three manuals (Assembly, Operator's and Trouble-shooting) in a hard-cover, 3 ring binder plus an Assembly Hints manual.

Buy our computer and we'll automatically make you a member of the *Altair User's Group*. You'll have access to a whole range of custom software designed exclusively for the *Altair 8800*.

We're quite serious about making computer power available to you at a price you can afford.

BASIC ALTAIR AND OPTIONS

The basic *Altair 8800 Computer* includes the CPU, front panel control board, front panel lights and switches, power supply and expander board (with room for 3 extra cards) all enclosed in a handsome, aluminum case.

Options now available include 4K dynamic memory cards, 1K static memory cards, parallel I/O cards, three serial I/O cards (TTL, RS232, and TTY), octal to binary computer terminal, 32 character alpha-numeric display terminal, ASCII keyboard, audio tape interface, floppy disc system, and expander cards.

PRICES: Altair Computer Kit with complete assembly
instructions **$439.00**
Assembled Altair Computer **$621.00**
1,000 word static memory cards   **$176.00** kit
                                                 & **$209.00** assembled.
4,000 word dynamic memory card **$264.00** kit
                                                 & **$338.00** assembled.

NOTE: Altair Computers come with complete documentation and operating instructions. Altair customers receive software and general computer information through free membership to the Altair User's Club. Software now available includes a resident assembler, system monitor, text editor, and Basic compiler.
Prices and specifications subject to change without notice. Warranty: 90 days on parts for kits and 90 days on parts and labor for assembled units.

MITS/6328 Linn N.E., Albuquerque, N.M., 87108, 505/265-7553

MAY 1975

Circle 16 on reader service card   **25**

Fig. 1.2 A 1975 ad for the Altair 8800. The Altair was the first personal microcomputer but it was designed following a much different approach than the KIM-1 and later machines. Note how the ad acknowledges the challenges of assembling the kit but emphasizes how rewarding the experience would be

MOS 6502 CPU as well.[1] Roughly at the same time, Tandy RadioShack went on introducing the TRS-80, completing a trio completing a trio that the computer magazine Byte nicknamed as the "1977 Trinity" and starting the personal/home computer era in grand style.[2]

As we can see, times were mature for personal and home computers to come and competition romped up quickly. The sudden interest and growing numbers of players in the personal computing space surely alarmed Jack Tramiel who, due to his previous negative experiences with typewriters and calculators, knew that, soon or later, someone would have come to conquer the market by bringing prices down.

This time, though, things were going to be different thanks to the MOS acquisition. Now Commodore had an advantage and Jack would have not allowed anyone, not even the Japanese, to undercut his newly found business niche.

Jack's next move became evident in an historical meeting hold near London in April 1980 where all top Commodore managers were gathered together to discuss upcoming strategies and new products. Over there Jack announced his intention to develop a new color computer to be sold at an extremely low price, less than $300, to finally start selling the computing revolution "to the masses, not the classes". Such a low price point, especially for a color computer, was completely unheard of but the initial shock and scepticism that such a plan raised around the table was quickly dismissed by Jack shouting "Gentlemen, the Japanese are coming, so we will become the Japanese!"[3]

This statement was actually a perfect synthesis of Jack's extremely aggressive approach to business. Especially after the previous defeats in the 60s and 70s, Jack

---

[1] The Apple I, also built around a 6502, was released as a single motherboard in mid 1976, for $666.66.

[2] The PET launch price was $795 while the Apple II was sold for $1,298 (computer only) and the TRS-80 retailed for $600 (including a monitor).

[3] Michael Tomczyk: "Home Computer Wars", Compute! Books, 1984.

Fig. 1.4 The VIC-20, first released as the VIC-1001 in Japan in late 1980 and then worldwide in 1981 with a price tag of $299.95. Byte magazine reviewed it in May 1981 with high praise: "even with some of its limitations [...] it makes an impressive showing against more expensive microcomputers like the Apple II, the Radio the Radio Shack TRS-80 and the Atari 800". This figure is licensed under CC and is credited as follows: Cbmeeks/processed by Pixel8—Original uploader was Cbmeeks at en.wikipedia

had no fear of potentially cannibalizing existing Commodore products and always pushed the company to compete with itself to ultimately improve its offerings: if they didn't do so, external competitors, whether Americans, Europeans or Japanese, would easily step in with far worse consequences for the company in the long run.

The low cost color computer idea took shape in just a matter of months in the form of the VIC-20 (Fig. 1.4), developed and marketed under the direct supervision of a newly hired manager named Michael Tomczyk.

To market the new product, Michael decided to present it as the "Friendly Computer". While trying to "humanize" computers wasn't a completely novel approach (Atari was proposing their line of computers as the "Computers for People" and the TRS-80 manual was trying hard to make users comfortable by avoiding overly technical concepts and jargon while stressing ease of use instead), it was the very first time the whole marketing campaign of a newly launched computer was entirely based around the concept of "friendliness".

To attract a new crowd of users and show how "friendly" the computer actually was, the marketing also emphasized the gaming qualities of the VIC-20, putting it in direct competition with home gaming consoles like the Atari VCS and Mattel Intellivision. Indeed, games were a very important component of the VIC and were developed not only by the group directly managed by Michael but also by a very young team of developers from a start-up named HAL Laboratory[4] in Japan. It was there, in fact, that the computer was first launched in October 1980 to test the market while, hopefully, also impress and scare off Japanese tech companies by showing that an inexpensive and good computer had already been done. The new home computer market, at least in the West, was going to be in Commodore's steady hands.

---

[4] HAL Laboratory was also going to develop several of the early C64 games like Jupiter Lander, Avenger and Le Mans. Later it went on to become a major console developer closely tied to Nintendo and responsible for great games like Kirby and many others. Their first VIC games were unofficial ports of famous arcade games like Space Invaders and Rally-X and showed the possibilities of the VIC-20 while also bringing in some legal trouble for Commodore.

Fig. 1.5 One of the original VIC-20 ads featuring actor William Shatner

The friendly and game focused marketing plan was then pushed forward masterfully by signing actor William Shatner, Captain Kirk of Star Trek fame, for a series of commercials, both printed (Fig. 1.5) and aired on TV.

Fig. 1.6 Jack Tramiel (*left*) and Michael Tomczyk celebrating the one millionth VIC-20 sold. The VIC-20 achieved that goal by the end of 1982: 800,000 units were sold in that year alone, a really impressive number if we think that the Apple II, released in 1977 released in 1977, had sold 700,000 units overall by then while sales for the newly launched IBM PC, between August and December 1981, were just around 13,000 units (photo by kind permission of Mr. Michael Tomczyk)

By selling via general retailers like Kmart and not by relying exclusively on specialized computer stores, the VIC-20 had no problems in reaching a new audience and was a resounding success: it was the first computer ever to sell one million units (Fig. 1.6) and, once discontinued in January 1985, sales were in excess of two and half millions.

Anyway, despite such an achievement, Commodore was not going to rest on its laurels and, true to Jack's golden rule of keep competing with yourself before others do, something even more impressive was already looming on the horizon: the Commodore 64.

Chapter 2
The Commodore 64 and Its Architecture

Abstract Here the unveiling of the C64 is discussed together with its original architecture, introducing the main components that made it unique: the 6510 CPU, the VIC-II and SID chips.

Keywords CES · Vertical integration · Computer architecture · 6510 · VIC-II · SID

Plans for the new machine that would ultimately become the Commodore 64 started hastily already in early 1981 and, by successfully meeting the seemingly impossible deadlines imposed by Jack, Commodore engineers were able to showcase the new computer prototype at the winter Consumer Electronic Show (CES) in Las Vegas in January 1982, alongside all the new material developed for the VIC-20, including the 1600 VICModem, the first inexpensive modem device that was also going to play a pivotal role in popularizing online activities in the years to come, as we will see in Chap. 7.

The showing of the new machine, which could use existing VIC-20 peripherals, was completely unexpected (indeed, not even many people at Commodore itself were aware of its development which was kept under the uttermost secrecy by Jack) and made a terrific impression thanks to several demos emphasizing its high end specs which included 64 KB of overall memory (direct competitors at the time had at most 48 KB), screen resolution able to comfortably display 40 columns of text (the VIC-20 could only display 22, making it unsuitable for serious word processing applications) and advanced multimedia qualities.

In August 1982 the C64 was ready for its commercial debut retailing at $595, a price tag that was thought to be impossible by the competition when originally announced at CES. Indeed, such a low price was only possible thanks to Jack's foresight in implementing a vertical integration manufacturing process centered around the original MOS acquisition done years earlier and it actually left Commodore with a very healthy profit, allowing the company to start pushing down costs even further soon after launch. This started a savage business war where Commodore could continuously slash retail prices to undercut competition

Fig. 2.1 The original Commodore 64 "breadbin" model in all its glory

while remaining profitable. The C64 (Fig. 2.1) was soon much cheaper than any other competitor, allowing Commodore not only to drastically increase its user base month after month but also to push many other home computer manufacturers into serious trouble or even force them entirely out of business like in the case of Texas Instruments and its TI-99/4A computer.

With worldwide distribution starting in early 1983, the C64 was an instant and ever lasting success that was discontinued only in 1994 once Commodore had to file for bankruptcy.

Overall, according to Commodore's 1993 Annual Report, the C64 sold 17 million units across its different iterations, like the C64C (an early example of what would be seen today as a "slim" hardware revision), SX-64 (a self contained and portable version of the computer), C64G (a later, cheaper breadbin version) and the C64GS (a C64 turned into a game console by removing the keyboard).

The Commodore 64 was a machine that refused to die and be forgotten, imposing itself as an 8 bit icon representative of a whole technological era. As we will see throughout the book, it affected the history and development of games and personal computing in every imaginable way across the years, inspiring people to innovate in several different areas.

But how did it do that? To understand its long lasting influence and how it managed to accomplish all this, we should start by having a quick look at its architecture and see what made it unique.

In its simplest form, like all modern computers, the C64 can be described in terms of the so-called Von Neumann architecture (Fig. 2.2), named after the mathematician John von Neumann (1903–1957).

As we can see from the diagram, a computer can be subdivided into three main parts:

- a **Memory**, to store data and instructions.
- **Input** and **Output** mechanisms to interact with the outside world.

Fig. 2.2 A schematic
diagram outlining the
components of the Von
Neumann architecture

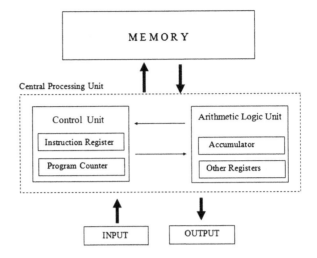

- a **Central Processing Unit** (CPU), whose main components are the **Arithmetic Logic Unit** (ALU), able to perform basic arithmetic and logic operations on data which are retrieved from memory and stored in a few dedicated registers for manipulation (the most important of which is traditionally called "Accumulator") and a **Control Unit** (CU) which implements the instruction set of the CPU and includes two special registers named **Instruction Register** (IR) and **Program Counter** (PC) to keep track of the instruction being executed and the one to execute next.

All of these components are then connected with each other via a bus system that allows data to be transmitted back and forth as needed.

Regardless of the model and origin, all CPU activities can be summarized as a sequence consisting of the following phases:

- **Fetch**: where an instruction stored in the memory location specified by the Program Counter is retrieved and stored in the Instruction Register. When done, the Program Counter is also updated with the memory address of the next instruction to load once the current one has been executed.
- **Decode**: where the Control Unit, holding the fetched instruction in its Instruction Register, dissects it to understand what needs to be done.
- **Execute**: where the CPU actually executes all the steps required by the specific instruction, for example by loading numbers into the various ALU registers to perform different operations, like an addition, whose result is then stored temporarily in the Accumulator for later use by a following operation.
- **Writeback**: where, as the name implies, the final result of the instruction is written back into a memory location or sent to an output device. After this, a new a fetch phase starts to load the next instruction and the whole sequence is repeated till the whole program has been executed.

The CPU mounted in the Commodore 64 was a 6510 (Fig. 2.3), a variation of the original 6502 which allowed for different configurations of the memory layout and

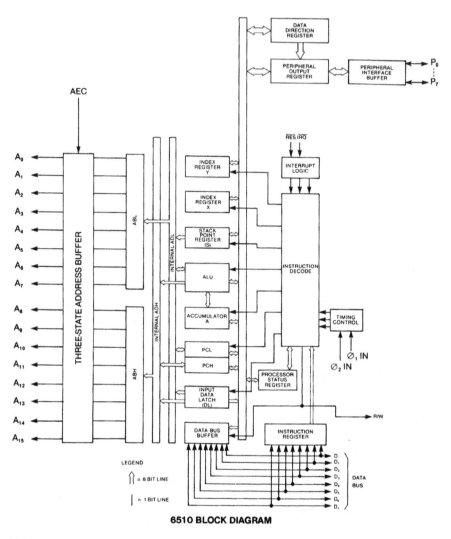

Fig. 2.3 The 6510 CPU block diagram where we can appreciate all its different components

for adopting a "tri-state" address bus, a modification useful for connecting or disconnecting different circuits. This feature was needed, for example, for granting the video chip direct access to the memory, disconnecting the RAM from the CPU itself.

A flexible and enhanced CPU like the 6510 was paramount to the success of the C64 but was not the only reason the new machine raised so much interest at the original unveiling during the 1982 CES, nor of its everlasting success. In fact, two other MOS custom chips were included to set the C64 apart from the rest of the competition: the Video Interface Controller (VIC-II) and the Sound Interface Device (SID).

To really appreciate the uniqueness of the machine, we should gain a little more understanding of what made these components so special.

2.1 The VIC-II

The follow up to the VIC-20's own VIC chip, the VIC-II[1] was designed primarily by Al Charpentier and Charles Winterble by taking into consideration all the strong points of existing graphic chips in competing machines, like native sprite support from the TI-99/4A or collision detection from Mattel's Intellivision, and then synthesize a new, superior one.

In the end, about 75 % of the VIC-II surface was dedicated to implement sprites related features. The chip also allowed for smooth scrolling across a screen resolution of up to 320 × 200 pixels where different colors, from a fixed palette of 16, could be displayed at once. As we see, arcade quality graphics for games were clearly planned from the very beginning of the design stage, with sprites, referred to in the technical documentation as "Movable Object Blocks" (MOB), being a top priority.

Programming wise, working with the VIC-II chip essentially revolved around manipulating its unique 47 registers, which handled all the information related to sprites X and Y coordinates on the screen, their colors, background and border colors, collision detection, the raster counter and more.

It is very interesting to note that, while the chip by default allowed for up to 8 sprites (each being 24 × 21 pixels if monochrome, 12 × 21 pixels otherwise) to be easily handled concurrently per scanline, this limitation could actually be overcome by a smart use of the provided raster interrupt routines, allowing for many more sprites to be displayed on the same screen at once. This technique, usually referred to as "Sprite Multiplexing", worked by triggering an interrupt to pause the drawing process when a given scan line was reached and then call a routine to modify and reload the VIC-II chip's registers as needed before proceeding in drawing the remaining part of the screen: by doing so, already drawn sprites could be relocated and drawn again in a different area of the screen, colors could be added, graphics modes changed and so on, allowing for a much richer and varied experience that could have been impossible otherwise (Fig. 2.4). Theoretically, with each sprite needing 64 bytes of memory to be stored and having the VIC-II 16 KB of overall addressable space for screen, character and sprite memory, up to 256 sprites could be displayed on a C64 screen!

Anyway, adding sprites was not the only trick possible and, as developers got more and more knowledgeable about all the different VIC-II nuances, understanding also how these could be manipulated by using specific quirks of different TV standards like PAL and NTSC, very impressive graphical effects like color

---

[1] IC Numbers: NTSC: MOS 6567/8562/8564, PAL: MOS 6569/8565/8566.

Fig. 2.4 Katakis, a graphically impressive space shooter released in 1989 by Rainbow Arts, used sprite multiplexing to the maximum effect, displaying many different sprites at once to create a frantic action experience

Fig. 2.5 Mayhem in Monsterland. Released in 1993 by Apex Computer Productions, Mayhem was one of the very last titles released on the C64. It was also one of the most technically impressive ever produced thanks to very advanced graphics enhancements, including some that were only possible on PAL systems, like PAL-colorblending, and made the game look almost like a 16-bit production

blending and color interlace could be achieved to show extremely colorful graphics and characters that were hard to match on any other 8 bit computer or gaming system (Fig. 2.5).

2.2 The SID

The SID was, arguably, the most beloved chip in the C64 and it is fondly remembered by many even today.[2] What made it so special?

---

[2] IC Numbers: MOS 6581/6582/8580.

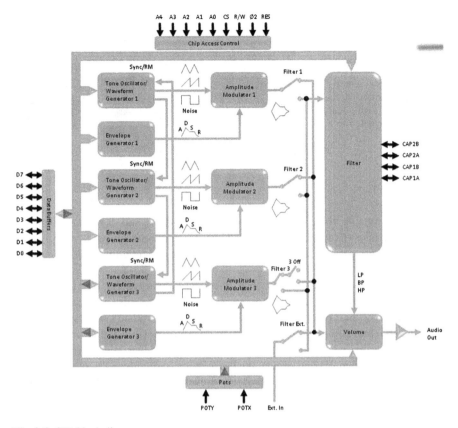

Fig. 2.6 SID block diagram

Designed by Robert Yannes, it was a very advanced sound synthesizer for the time and the first sound chip sporting an envelope generator to be integrated in a home computer.

In particular, the SID made available three synthesizer voices that could be used independently or in conjunction with each other to create complex sounds. Each voice consisted of a Tone Oscillator/Waveform Generator, an Envelope Generator and an Amplitude Modulator (see Fig. 2.6). Manipulating the tone oscillator allowed for fine control of the pitch across a very wide frequency range (8 octaves, approximately from 16 to 4,000 Hz) while the amplitude modulator of the oscillator could dynamically control the volume of the sound according to the input from the envelope generator. The latter could be programmed by specifying different attack/decay/sustain/release (ADSR) values to shape the volume of a note or sound effect in any way the sound designer desired.[3]

---

[3] More on this in Chap. 3.

Fig. 2.7 Paul Norman's
Forbidden Forest [(c)1983
Cosmi] was one of the first
titles to push the SID unique
capabilities and showcase
a dynamic musical score
able to follow and haunt the
player, creating a truly scary
and eerie atmosphere for the
game

A programmable filter featuring low-pass, high-pass and band-pass outputs, with 6 dB/octave (band-pass) or 12 dB/octave (low-pass/high-pass) rolloff, was also available for generating even more complex and dynamic tone colors via subtractive sound synthesis.

Last but not least, the SID also had two A/D converters for interfacing with potentiometers and could even process external audio signals, allowing multiple SIDs to be daisy chained or mixed in complex electronic music systems.

All these features together allowed for tremendous flexibility that, in the hands of talented composers and audio programmers like Ben Daglish, Martin Galway, Rob Hubbard, Jeroen Tel and others, pushed gaming audio to new heights: original sound effects and engaging music that could even change dynamically following the gaming action, like in Cosmi's 1983 hit "Forbidden Forest" (Fig. 2.7), were finally possible and were a huge step forward compared to anything done before.

Many exceptional sound tracks followed the pioneering efforts of Forbidden Forest, with music constantly gaining more relevance and importance across game development.

Games like Parallax (music by Martin Galway), Delta (music by Rob Hubbard), the Last Ninja (music by Ben Daglish) and Supremacy (music by Jeroen Tel), among others, had music tracks that set them apart from anything else and are still listened to and enjoyed even today, not only when playing the actual games but also via standalone SID playing utilities and emulators.[4]

Now that we have a basic understanding of the C64 main hardware components, we can progress to learn about its actual operation and programming.

---

[4] Interested readers may check the freely downloadable Java SID Player Music Library: http://sourceforge.net/projects/jsidplay2/.
An online HTML5 emulator is also available at http://www.wothke.ch/experimental/TinyJsSid.html.

Chapter 3
Ready

Abstract This chapter introduced the reader to the joys and sorrows of the C64 original OS and its BASIC programming language, including a foundational knowledge of sprites and sound programming via the VIC-II and SID chips respectively.

Keywords Microsoft BASIC · Programming

Arguably, one of the most iconic images associated with the Commodore 64 is its booting screen (Fig. 3.1).

"Ready". Period.

But... Ready for what? This is what countless neophytes asked themselves the very first time they switched on their computer. Such a simple line followed by a blinking cursor had surprising and far reaching implications: expecting users to start typing something, maybe a command for listing files stored in a floppy disk, loading a game or, possibly, take the first step in developing a complex program or great new game, gave young users the feeling of being in control with an apparently limitless range of possibilities at their fingertips, exciting their imagination. Many felt the C64 willingness or, we should say, "readiness", to execute their commands was an implicit invite to find out more and venture into a new, uncharted territory. They started learning programming on their own, often proceeding by trial and error with only the good but limited support provided by the official Commodore User's Guide, possibly sharing their experiences with a small groups of like-minded friends. Computer programming was, finally, within the reach of millions.

The soaring enthusiasm of this new generation of users even made them overlook the machine's objective deficiencies. Regrettably, the built-in OS and programming language was actually one of the C64 weakest features: the Commodore 64, in fact, shipped with a built-in version of Microsoft Basic 2.0 and, while Microsoft Basic was indeed an ubiquitous programming language at the time and variations of it appeared on all the most well known home computers, from the Atari 400/800 to the Apple II, Version 2.0 was pretty old and was definitely showing its age by mid 1982. It was the same version that shipped not only with the VIC-20 but also with the original PET 2001 series years earlier. It is not surprising then

© Springer Science+Business Media Singapore 2015
R. Dillon, *Ready*, DOI 10.1007/978-981-287-341-5_3

Fig. 3.1 The C64 booting screen: Ready to start!

to realize it lacked many improvements and useful functions that newer versions had. For example, variable names were limited to two characters only (i.e. two variables named "player1" and "player2" would actually be seen as the very same variable, named "pl", by the interpreter) and commands for simplifying file access were missing: while in the newer Microsoft Basic 4.0 to delete a file on a floppy disk it was enough to write the command SCRATCH followed by the file name, Basic 2.0 users had to write something much more convoluted to let the C64 interact with the DOS installed on the disk drive in a step-by-step manner. For example:

OPEN 1,8,15
PRINT #1, "SCRATCH:filename"
CLOSE 1

Let's analyze each line of this short code snippet to understand how the OS worked: the first line opens a file channel for communication between the C64 and the disk drive, making the latter ready for use. Specifically, the first parameter is an arbitrary *file channel number* between 1 and 254, defined by the user. The second identifies the device (the floppy drive in this case) while the last parameter specifies a secondary address for the channel, ranging between 0 and 14 for data transfer or 15 for issuing commands, like in this case.

The second line, via the PRINT#1 instruction, directs (i.e. "prints out") the upcoming command to the file channel opened previously. Here we are here issuing the "SCRATCH" command followed by the actual file name.

The file channel needs then to be closed explicitly by the user via the CLOSE command.

Quite troublesome and, unfortunately, file operations weren't the only area where high level instructions were missing.

Extensions for easily handling the advanced graphical and sound capabilities that made the VIC-II and SID chips unique were also unavailable from Basic.

Luckily, though, this didn't seem to have discouraged many young programmers: on the contrary, having experienced those constraints and being forced to work at a lower level, instructing the machine step-by-step even for tasks as simple as deleting a file, may have actually helped aspiring developers in the long run. Having gained a better understanding of computer operations by surviving a rougher start, could have in fact eased the passage to more advanced programming later on, establishing the right mindset for writing complex application and arcade quality games via an assembler, for example.

Regardless, it was still possible to use the VIC-II and SID features from a Basic program but handling sound and graphics required the use of the infamous PEEK and POKE instructions.

These were commands that allowed to check a value ("peek") or inserting one ("poke") in any memory location of the computer. Needless to say, an extensive and detailed knowledge of how the memory was mapped was essential: "poking" a value in a wrong location would easily result in unpredictable behaviours and system crashes.

Regarding the VIC-II, all the different registers we talked about in Chap. 2 were mapped between memory addresses 53248 and 53294 ($D000 and $D02E in hexadecimal)[1] with the VIC assuming X and Y coordinates of its eight sprites were inputted by the programmer at the beginning of that area, i.e. 53248 and 53249 store X and Y coordinates for Sprite 1 respectively, 53250 and 53251 store coordinates for Sprite 2 and so on up to 53262 and 53263 for Sprite 7,[2] while the byte for enabling the various sprites on screen was set at location 53269 ($D015) where each bit represented a sprite: setting it to 1 would switch the corresponding sprite on and make it visible on screen.

But where should the actual sprites be stored in memory?

The 64 kB overall memory of the C64 was divided into four contiguous banks of 16 kB each and the VIC-II could only address one of these at a given time. Which bank was chosen and where exactly in this bank the graphical data were actually going to be stored was determined on a case by case basis by the programmer. Needless to say, this area had to be chosen very carefully to avoid overlapping with any sensitive data.

To direct the VIC to read sprite data from a particular area in the assigned memory bank, memory locations between 2040 ($07F8, for sprite 0) to 2047 ($07FF, for sprite 7) were reserved to work as pointers. Here the programmer would store the actual memory address to finally tell the VIC where to look for each sprite.

---

[1] See Appendix C for a comprehensive list of each VIC-II register, its memory location and purpose.

[2] A particularly attentive reader may have noticed that, while the C64 has a standard screen resolution of 320×200 pixels, for storing a sprite's X coordinate we are only using a specific memory location worth 1 byte of memory, i.e. 8 bits, which can cover values from 0 to 255 only! How do we place a sprite having a X coordinate between 256 and 320 then? Indeed, for each sprite, a 9th bit is needed to represent such values and it is stored in location 53264 ($D010) where each one of its 8 bits, from 0 to 7, is being assigned to the corresponding sprite.

Fig. 3.2 Designing a sprite
on a 24 × 21 grid

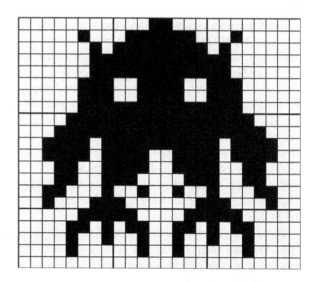

Clearly, assimilating all these steps was not easy and, most importantly, they had to be handled rigorously and precisely. A practical example would help in understanding the workflow and mindset that early programmers had to adopt when programming the VIC chip. It is then useful to start by discussing how sprites were actually represented and how much memory each of them took.

We already saw in Chap. 2 that each monochrome sprite would take 24 × 21 pixels. For example, if we were designing a sprite for some sort of monster, we would then start by drawing it in a 24 × 21 grid like this:

Notice how each row can be subdivided into three groups of eight cells each. With each cell being a bit, with a value of 1 when drawn or 0 otherwise, each row can be represented by three bytes and a corresponding binary value can easily be computed to store it: for example, the first row of Fig. 3.2 can be described as 00000000, 00011000, 00000000 or, in decimal 0, 24, 0. Having 21 rows means 63 bytes are needed to store the whole sprite data.

In Basic 2.0 we could simply start writing a new program by typing in instructions on numbered lines and then press "Return" to confirm them. In this case, we would begin by initializing a couple of variables to hold the first VIC memory location and the desired memory location we will be using to store our monster sprite data (e.g. 15872, i.e. $3E00):

1   VIC = 53248 : REM the VIC chip starts here[3]
10   LOC = 15872 : REM we will store our sprite data from this location onwards

This would then be followed by a POKE instructions to notify the VIC about which specific sprite, among the eight possible ones, was going to be used and where its data would be stored:

---

[3] Comments in MS Basic could be added by using a REM statement. Using a colon instead allowed for concatenating two or more instructions on the same line.

20   POKE VIC+21,4 : REM switching on sprite number 2
30   POKE 2042,248 : REM pointer to Sprite 2 memory bank

Line 20 writes value 4 (corresponding to 00000100 in binary) in location 53269, i.e. memory location 53248+21. That is the VIC registry that keeps track of which sprites are active at any given time. With this instruction we are now turning the third bit on so, in this case, we will be using the third sprite (or sprite number 2, if we see sprites as numbered from 0 to 7, like the C64 does).

The next POKE writes value 248 in the location working as a pointer for Sprite 2. Why 248 and not 15872? Don't forget we are working with an 8-bit machine here so the maximum number we can store in an individual memory location is 255 only! 248 is then multiplied by the computer by 64 (since each sprite needs 63 bytes of data) so that the VIC can actually start looking for the sprite data at location 248 * 64, i.e. exactly the 15872 we wrote earlier.

We can now actually save the sprite data in memory. To do so, we can use a FOR...NEXT cycle to easily input our 63 bytes in the memory we assigned them:

40   FOR N = 0 TO 62 : REM we have to read 63 bytes to represent the shape of a sprite
50   READ D : REM Basic will automatically look for a data block to read from
60   POKE LOC+N, D : REM each byte is stored in its corresponding location
70   NEXT N
110   REM sprite data starts here. We write 3 bytes per line for easy comparison with Fig. 3.2
120   DATA 0,24,0
121   DATA 4,126,32
122   DATA 2,255,64
123   DATA 3,255,192
124   DATA 3,255,192
125   DATA 7,60,224
126   DATA 7,60,224
127   DATA 7,255,224
128   DATA 15,255,240
129   DATA 31,255,248
130   DATA 61,255,188
131   DATA 61,231,188
132   DATA 25,231,152
133   DATA 25,195,152
134   DATA 49,36,140
135   DATA 35,129,196
136   DATA 7,195,224
137   DATA 13,102,176
138   DATA 9,36,144
139   DATA 8,36,16
140   DATA 0,0,0

Besides the data defining the shape of our sprite we still have to tell the C64 other details before finally being able to display our work on screen like, for example, the sprite color and its actual location.

By knowing that, for each sprite, colors are saved into memory locations between 53287 and 53294 while X and Y coordinates have to be stored between 53248 (which is the first VIC-II location we already assigned to variable VIC in line 1) and 53263 respectively, we can proceed by adding the following instructions[4]:

80   SC = 53287 : REM first location used to store colors. This one refers to Sprite 0
81   POKE SC+2,7 : REM since we are using Sprite 2, we need to poke the color code in 53289. 7 means yellow
85   POKE VIC+4, 170 : REM VIC+4 identify the X coordinate for Sprite 2. We defined VIC in line 1
86   FOR Y = 0 TO 240 : REM using a FOR...NEXT cycle for implementing vertical movement
87   POKE VIC+5, Y : REM every frame we move the Sprite 2 down one pixel
88   NEXT Y
90   GOTO 86 : REM reset the Y position and repeat the scrolling till the program is stopped.

If we were now to type and run this small program into a real C64 or an emulator,[5] we would then see the sprite designed in Fig. 3.2, colored in yellow and moving from the top to the bottom of the screen.[6] On the other hand, if we wanted to control the sprite via a joystick, the constant check via a PEEK instruction of location 56320 (for the joystick in port 2, for example) would have been required: a value of 127 meant no input while 125 meant up, 126 down and so on. Once movement is detected, the programmer could then proceed in updating the sprite coordinates accordingly.

Moving from programming graphics to working with audio meant dealing with the SID chip and all its corresponding registers instead. These, like those of the VIC-II, were memory mapped to the C64 RAM itself, specifically between locations 54272 ($D400) and 54300 ($D41C).[7] Likewise, learning audio programming techniques wasn't easy at first but would prove extremely rewarding to those who would ultimately master its nuances.

In any case, regardless of the final result and its complexity, all sound generation revolved around the definition and manipulation of a few fundamental

---

[4] Note that, to keep the overall program more tidy and readable, we are inserting the new lines before the data block by numbering them accordingly.

[5] Check Appendix B for a list of emulators and other programming tools.

[6] For placing sprites in specific positions, we must be aware that, while the inner C64 display window is 320 × 200 pixels as discussed, sprites can also be placed also under the external borders, making the overall area for sprite movement 368 × 255 pixels. Location (0, 0) then is not at the upper left corner of the 320 × 200 window as we might have expected, but at the top left of the outside border.

[7] See Appendix D for a reference.

Fig. 3.3 A typical attack, decay, sustain, release (*ADSR*) envelope used to simulate more natural sounding notes and effects. All these parameters can be controlled independently for each note by the SID via its registers

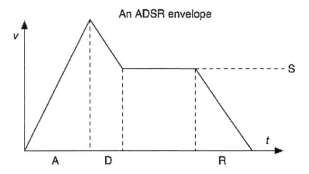

parameters that are standard in all sound synthesis systems: peak and sustain volume of the sound, its attack, decay, release and overall duration (to define the sound's own ADSR envelope, see Fig. 3.3), its waveform type and the actual frequency to be played.

A simple code snippet to play a note with a timbre sounding more or less like a positive organ would then look like the following:

```
1    FOR CL = 54272 TO 54296: POKE CL,0: NEXT : REM resets all SIDSID
     registers
10   POKE 54296,15 : REM volume
20   POKE 54277,100 : REM attack/decay
30   POKE 54278,215 : REM sustain volume/release time
40   POKE 54273,16: POKE 54272,195 : REM frequency
50   POKE 54276,17 : REM selects the waveform and starts it
60   FOR T = 1 TO 1000 : NEXT : REM wait for a while: a simple way of defin-
     ing the sustain time
70   POKE 54276,16 : REM stops the note, starting its release phase
```

To understand what we are actually poking into the SID registers to generate the sound, let's analyze each line in detail.

After resetting every SID register in line 1 (always a good habit), line 10 sets the peak volume our note will have at the end of the Attack phase. Note though that only the first nybble, i.e. the first 4 bits (bit 0–3), of register 54296 are dedicated to the sound loudness,[8] meaning that we can only have a range of values between 0 to 15 (with 15 being the loudest).

Line 20 defines the duration of the Attack and Decay phases. These are stored in location 54277 with a nybble dedicated to each: bits 0–3 are assigned to the Decay while bits 4–7 to the Attack. The possible 0–15 range is then mapped to cover a huge interval of possible time values: from 6 ms (for 0, i.e. bits set as 0000) to 24 s (for 15, i.e. bits set to 1111) for the Decay and from 2 ms to 8 s for the Attack (i.e. bits set to 0000 or 1111 respectively). In the specific example, line 20 inputs

---

[8] The second nybble is used to enable filtering.

the decimal value of 100, i.e. 0110 0100 in binary. Checking the look up table shows an Attack value of 6 corresponds to 68 ms, while a Decay value of 4 corresponds to 114 ms.[9]

We take care of the Sustain volume and Release time in line 30 in a similar fashion. Bits 4–7 are those dedicated to the volume, again giving us a 0–15 range of possible loudness values to play with. Bits 0–3 specify the release time instead, following the same approach seen for defining Attack and Decay times: a value of 0 will set a release as abrupt as 6 ms only while a value of 15 will make the sound slowly fade away for as many as 24 s!

In this case we are poking in a value of 215, which is in 1101 0111 in binary, i.e. we are setting the Sustain volume to 13 and the Release time to 7, which equals 240 ms.

Specifying the actual note pitch takes two bytes, since the SID could generate sounds across a wide range of frequencies covering eight octaves. In line 40 we are instructing the SID to play note 16 195, which corresponds to a C in the fourth octave.

The bits in the register mapped at location 54276 has several functions. In line 50 we are switching to 1 bits 0, to start the note, as well as bit 4 to select the triangle waveform. Turning on bits 5 or 6 would have selected the saw or square waveforms instead.

Line 60 is a simple FOR...NEXT cycle to make the computer wait for a little while and extend the sustain time. Finally, in line 70 we go back to register 54276, turning bit 0 to 0 but leaving bit 4 to 1, effectively stopping the sustain so that the note can progress to its release phase and then die off naturally.

Programming sound effects worked in a similar way and involved a lot of experimenting with different settings and values to literally sculpt the sound in the way desired.

For example, a simple SFX for a gunshot could be programmed by selecting the noise channel (bit 7 in 54276), setting the highest possible volume, defining very fast Attack and Decay times and then set a Release time possibly in accordance with the reverberation suggested by the particular virtual environment:

```
10   POKE 54273,21: POKE 54272,31 : REM set a low frequency for the gunshot
20   POKE 54296,15 : REM volume maxed
30   POKE 54277,16 : REM attack/decay set to 8 and 6 ms respectively
40   POKE 54278,250 : REM setting sustain volume to 15 and Release time to 1.5 s
50   POKE 54276,129 : REM select noise channel. Start note (Attack, Decay, Sustain)
60   POKE 54276,128 : REM stop the Sustain to start the Release phase and end
     the SFX
```

But how come the C64 had all these OS and programming shortcomings in the first place when the engineers who designed the machine and its functionalities were among the brightest in the business?

There were several reasons.

---

[9] See Appendix D for the different correspondences between numerical values and timings.

First of all, engineers were under tremendous pressure from Jack for delivering the quickest implementation possible. They needed something that worked reliably even if it wasn't exactly state of the art so fixing these issues was not considered high priority.

Arguments with marketing were also frequent: marketing director Kit Spencer was aware that new business computers were adopting specific operating systems, like CP/M, that were becoming more and more popular and strongly pushed for the C64 to adopt it as well. Unfortunately, doing so would have required additional resources and development time, both of which were lacking. In the end, a sort of compromise was reached: marketing had to acknowledge that implementing CP/M in ROM wasn't feasible while the engineering team also acknowledged that having CP/M would indeed effectively improve the C64 chances of success as a business machine. It was then decided to include it later as an add-on cartridge, leaving the C64 with its weak OS for the time being.

Regrettably, when such a product consisting of a combination of disk and cartridge integrating a Z-80 CPU (see Fig. 3.4) was actually released a few months after the C64 launch, the implementation suffered of poor performance besides a lack of actual disk compatibility between native CP/M drives and the Commodore's 1541 disk drive, which was particularly slow since it had to be designed by taking into account the VIC-20 needs as well. This was a real issue since it required that, to run on a C64, all CP/M software had to be rewritten to make it actually readable by the 1541 drive first! In the end, very little of the huge

Fig. 3.4 Commodore's implementation of the CP/M was a brave but ineffective attempt to make the C64 more appealing to business users. Unfortunately it fell short of expectations mainly due to media incompatibilities between disk formats, meaning that no existing CP/M software packages could have been used out of the box on a C64 but had to be rebuilt taking the differences into account

Fig. 3.5 The super expander cartridge, adding new basic commands for a more straightforward handling of graphics and music. A C64 without a cart couldn't run the resulting programs, though

software library originally released for the CP/M was made available and the CP/M cart, as predictable, had no commercial success and was soon forgotten.[10]

The C64 had no choice but to keep going with its original OS despite all its limitations, at least till third party improvements, like **JiffyDOS**, were released or new operating systems experimenting with windows and icons gave it a more professional look and feel, as we will see in Chap. 6.

Regarding Basic, the development team faced similar time related issues and, while the C64 was definitely planned to be a powerful games machine from the ground up, commercial quality games were never meant to be programmed in Basic anyway, meaning that easy to use instructions to access the VIC and SID chips were also not considered to be high priority. In any case, more powerful implementations of the language soon surfaced, like **Simons' Basic**, developed in 1983 by 16 year old British programmer David Simons and featuring a wide range of 114 additional commands, or the **Super Expander** cartridge, a less comprehensive extension that mostly focused on graphics and sound. Both were officially released by Commodore in 1983 and made Basic programmers' lives much easier (Fig. 3.5).

---

[10] A better implementation of CP/M was later built in the Commodore 128. This didn't help the new computer either, though, since, by that time, CP/M was past its prime and MS-DOS was rapidly becoming the leading business OS.

Chapter 4
Games, Games and More Games!

Abstract Critical analysis, by genre, of more than 100 games on the C64 that defined 8 bit gaming. The following genres are covered (in alphabetical order): Action Adventures, Adventures, Arcade conversions, Driving games, Edutainment, Movie Tie-ins, Platformers, Puzzles, Role Playing Games, Shoot 'em ups, Sports, Strategy, Virtual life, 3D Games.

Keywords Evolution of games · Game history

Despite Commodore efforts to propose the C64 as an all-rounder able to stand its ground also as a business computer, it is undeniable that its first and foremost use, and strength, was in entertainment and, more specifically, in games.

More than 10,000 games were officially published during the C64 lifetime by a new generation of young and passionate developers. While many of these were non exclusive but shared across other competing platforms as well, they often originated or found their best rendition on the C64, making it the platform of choice for many gamers and developers alike.

The sheer amount of creativity showcased in the games made by this extremely energetic and lively community can't be underestimated and, as we are going to see in the sections that follow, the C64 game scene influenced electronic entertainment as much as, if not even more, any dedicated gaming console that came before and after it.

To have an idea of the accomplishments made during these pioneering, but by no means primitive years, some among the most significant titles across several genres will be discussed, starting with Action Adventure games and then proceeding, in alphabetical order, with Adventures, Arcade conversions, Driving games, Edutainment, Movie tie-ins, Platformers, Puzzles, Role Playing Games, Shoot 'em ups, Sports, Strategy and Virtual life, to end with vector and polygon based 3D Games.

4.1 Action Adventures

Any innovative and technically impressive action adventure game so popular today, able to immerse players in new worlds thanks to a mix of real time exploration, combat and environmental puzzles, can be traced back to the early eighties when they soon became a trademark of 8 bit computers. On the C64, the genre evolved throughout the machine lifespan and gradually substituted more traditional and slow paced adventures in the hearts of many.

One of the first significant achievements in this area was "The Staff of Karnath" (Fig. 4.1). Developed by Ultimate in 1984, it was the first C64 exclusive title released by the same company that was revolutionizing the ZX Spectrum world thanks to masterpieces like "Atic Atac" and "Knight Lore". In "Staff of Karnath", as well as in its sequel, "Entombed" (1985), the player took the role of Sir Arthur Pendragon, here on a quest to retrieve eighteen pieces of a pentacle which were scattered across a castle infested by every imaginable kind of monster. Only once all the pieces were collected and brought over, one by one, to an obelisk hidden deep in the castle dungeons, Sir Arthur could finally gain access to the titular staff and destroy the evil within.

The side view perspective, while far less impressive than the isometric approach offered via Ultimate's proprietary Filmation engine used in Knight Lore, was still a worthy achievement: backgrounds offered plenty of detail while big, multi-colored sprites enjoyed freedom of movement within the castle's walls. Different puzzles were also present, starting from the various spells available for which there was no explanation whatsoever and it was up to the player to find out what they did and how to use them effectively.

In any case, even more detailed, though monochromatic, worlds viewed from an isometric perspective were soon offered in games like "The Great Escape" (Fig. 4.2) and "Fairlight" (Fig. 4.3).

Set in 1942, the former was an open game taking place in a very well rendered German POW camp on the North Sea during World War II and was so impressive

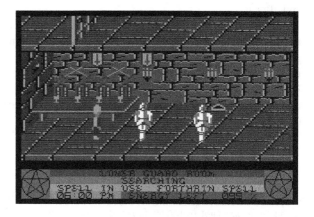

Fig. 4.1 The Staff of Karnath (© 1984 Ultimate). The beginning of the game, with Sir Arthur trying to reach a pentacle piece while avoiding two not-so-friendly knights

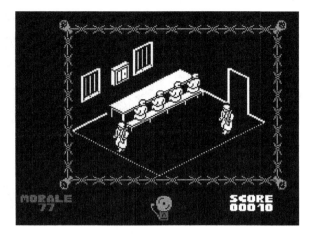

Fig. 4.2 The Great Escape (© 1986 Ocean). Breakfast time under strict guards surveillance. Notice how, instead of a common "energy" bar, here we have a "morale" counter. If the player is caught after missing a scheduled event, he will be confined in an isolation cell for a day, with a consequent morale drop. Once this reaches zero, the prisoner will forever lose hope to regain his freedom and surrender to his sad destiny: surely an original way for triggering the Game Over screen!

Fig. 4.3 Fairlight (© 1986 The Edge). Searching for the Book of Light in an adventure masterpiece originally created on the ZX Spectrum by Bo Jangeborg

to be touted by journalist Julian Rignall as "unquestionably the best arcade adventure so far this year".[1] Indeed, the attention to detail was staggering and this, together with multiple ways to plan and execute our escape (for example via excavating tunnels, using camouflage and so on) added a lot of room for experimentation with a consequent high replayability factor that made the game a favourite among both critics and players alike.

---

[1] Zzap!64, issue 27, July 1987, p. 104.

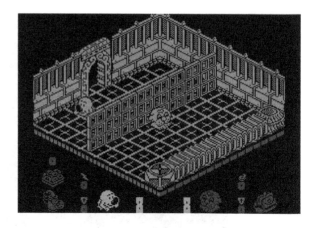

More traditional in its fantasy setting, Fairlight was also an interesting adventure that, despite its lack of speed, still managed to build an impressive atmosphere. Its highly interactive environments were particularly noteworthy: most furniture, like table and chairs, could in fact be pushed along and used not only to solve environmental puzzles and reach other objects but also to keep enemies at bay.

Anyway, isometric perspective adventures, where colors were sacrificed for allowing additional graphical detail in the various scenes, had their most successful representative in "Head Over Heels" (Fig. 4.4).

While being at first sight just another "escape the castle" type of adventure, the game quickly showed its uniqueness and originality by having the player switch control between Head and Heels, two distinct characters each with specific and complementary skills, all of which where needed to overcome several puzzles and enemies. Heels could run and carry objects while Head could throw paralyzing donuts. Separated at first, Head and Heels could later unite (with Head jumping over Heels, naturally) to join forces and ultimately escape and free the Blacktooth Empire together.

According to many, though, despite Head Over Heels' undeniable charm, beauty and success, the pinnacle in the action adventure genre on the C64 was achieved by another franchise: the Last Ninja, especially in its first two instalments (Figs. 4.5 and 4.6).

Released in 1987, "The Last Ninja" was an outstanding product under almost every aspect: graphics were as detailed as they were colorful, animations were smooth and realistic and the musical score was equally impressive. It was clear by now that experienced developers mastered the hardware to achieve results that would have been unthinkable just a few years earlier on the same computer.

The game saw a lone ninja survivor, Armakuni, in a quest to revenge his clan murdered by the evil Shogun Kunitoki and retrieve the legendary Koga scrolls. The adventure took place across 150 screens split among six areas of varied settings, ranging from open wastelands to castle dungeons to end in an inner sanctum temple. Combat involved only a couple of basic moves but different weapons, including throwing stars, a katana, nunchaku and more, still managed to keep it

Fig. 4.5 The Last Ninja (© 1987 System 3). With a budget of 300,000 British Pounds, the Last Ninja was one of the most expensive games ever produced back in the day. Indeed, production values were among the highest ever seen in a video game at that time

Fig. 4.6 The Last Ninja 2: back with a Vengeance (© 1988 System 3). From feudal Japan to XX century New York: the fight continues across even more stunning and colorful scenarios

fresh and interesting throughout the game. In the Last Ninja, every single element worked together to build a magical and somewhat stealthy atmosphere that was truly immersive and helped making the few flaws the game had forgivable: some pixel perfect jumping sequences were really frustrating and it was possible to miss some objects later required to solve a puzzle, making the game impossible to finish and forcing a restart. Still, despite these flaws, "The Last Ninja" quickly became one of the most successful and admired games on the C64, like its sequel, published 1 year later. There, the action was set in modern New York where Armakuni was teleported by the Gods to keep fighting Kunitoki, who managed to travel in time avoiding death after being defeated in the first game thanks to a magical orb.

Maintaining, and even surpassing, the very high standards set by the original game, Last Ninja 2 succeeded in delivering even more impressive and varied backdrop graphics, animations and, last but not least, also offered a limited edition including a ninja hood and a rubber shuriken for the joy of collectors.

Isometric perspective was not the only possible way to craft engaging action adventure games, though, and even an "old fashioned" flat 2D side view could still bring to life a very original concept, as demonstrated by "Samurai Warrior: the Battle of Usagi Yojimbo" (Fig. 4.7).

Fig. 4.7 "Samurai Warrior"
(© 1988 Firebird). It may be
a world populated by animals
but the bushido code still
holds in this adaptation of
Stan Sakai's "Usagi Yojimbo"
comic books. Following an
original design decision, the
game lacked a traditional
scoring system and instead
opted for a "karma" based
value that could rise or
decrease according to the
player's behaviour

Based on Stan Sakai's comics, the game was set in an alternative feudal Japan populated by animals where players took the role of the titular character, a ronin rabbit, in a quest to restore order to a land fallen into anarchy and free his friend being imprisoned by a rival Lord. Merging arcade adventure with hack and slash elements, the game emphasized a well crafted combat system which perfectly complemented the story progression, world exploration and the resulting interaction with many different NPCs. This was particularly original since it was based on a moral system (karma) that had to keep under control: while the player was theoretically free to behave more or less honourably across the various events in game, a proper overall behaviour was needed: for example, showing respect to villagers and travelling monks would make the karma rise while attacking and robbing them would decrease it. If the karma score would drop below zero, the hero would commit seppuku and kill himself, regardless of how much life energy was left or how many riches were accumulated.

A more advanced 2D side view with some perspective added to give a more 3D feeling and allow movement across the screen and access to different areas, effectively improving on the original "Staff of Karnath" template, was the choice in two of the last great arcade adventures on the C64: Tusker (Fig. 4.8) and Project Firestart (Fig. 4.9).

Clearly inspired by the figure of Indiana Jones, Tusker started off with the player somewhere in the Sahara desert looking for his father who went missing while searching for the mythical "elephant's graveyard". Plenty of enemies, in human, animal and undead form, and a constant risk of dehydration forced the player to act quickly and decisively across an engaging and treacherous world that spanned not only the initial desert but also jungles, villages and temples, with puzzles increasing in number and complexity as the game went by.

A very cinematic approach was at the heart of Electronic Arts' "Project Firestart" where, in a plot reminiscent of the movie "Aliens", we needed to infiltrate a spaceship where something went terribly wrong. Different weapons, good controls and an interesting storyline made this a memorable action adventure with

Fig. 4.8 Tusker (© 1989 System 3). Compared to Last Ninja, here System 3 ditched the fascinating but somewhat tricky isometric view for a simpler perspective that allowed a more intuitive control scheme

Fig. 4.9 Project Firestart (© 1989 Electronic Arts). Another space experiment involving aliens went terribly wrong. Are there any survivors? Can we find out what happened before it's too late?

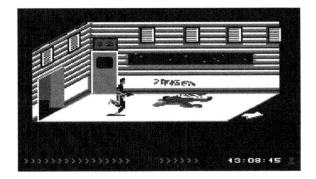

a survival horror atmosphere that, arguably, was matched only many years later by another Electronic Arts' title: "Dead Space" in 2008.

4.2 Adventures

Adventure games are one of the oldest genres, dating back to the days when mainframe computers could only display a few lines of text. Graphical capabilities of the early home and personal models were often very limited as well, making text based games the only suitable type of game to develop and play. The company that was responsible for popularizing the genre, as well as for setting very high quality standards that are still admired nowadays, was Infocom. Founded in 1979 by a group of MIT graduates, their games were published on almost every home computer available at the time and titles like the Zork trilogy (1980–1982), Planetfall (1983, the first computer game that made people cry) and Suspect (1984) are masterpieces that defined Interactive Fiction. These games are as engaging and immersive to play today as they were back in the day thanks to a very advanced parser that allowed for writing relatively complicated sentences in plain English.

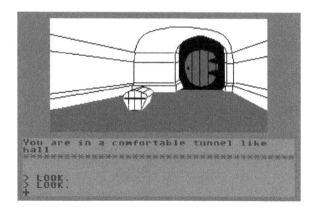

Fig. 4.10 The Hobbit (© 1983 Melbourne House): the first computer game based on J.R.R. Tolkien's classic story, Melbourne House's Hobbit was a very engaging rendition thanks also to colorful, albeit simple, graphics that were drawn in front of the player's eyes when entering new locations

Anyway, as soon as computers were able to display images, it was clear that most adventurers around the world were really longing for some visual feedback from the scenes that were described via the written text. On-Line Systems, soon to be renamed Sierra on-Line, was the first company to understand this need. After their first game featuring images to complement the in-game text sparked a lot of interest and sold beyond expectations ("Mystery House" for the Apple II, published in 1980), many other companies quickly followed, pushing adventure games towards new directions.

One of the first adventure games to showcase some inspiring artwork on the C64 was "The Hobbit", based on Tolkien's classic novel (Fig. 4.10).

The trend continued and, while most games adding some "eye candy" images were far from the quality of Infocom's text-only adventures and could only understand basic verb plus noun combinations, it wasn't long before some games appeared showcasing both an advanced parser as well as stunning images. This was the case for Magnetic Scrolls' "The Pawn" (Fig. 4.11) and its spiritual sequel "Guild of Thieves" (1987).

The Pawn was set in the fantasy land of Kerovnia, where for some mysterious reason, the player was suddenly teleported into. The set-up was quite elaborated and a novella included in the package was required reading to understand the net

Fig. 4.11 The Pawn (© 1985 Rainbird/Magnetic Scrolls): how did we get here? It will be a long journey to unravel all the secrets of Kerovnia!

of political intrigue and struggles that were happening and that, now, involved the player as well. Complex characters, humour and an intriguing story made the game one of the highest achievements in the adventure genre and, likely, also the swan song of text based games. It was soon clear, in fact, that building an advanced parser to compete with Infocom or Magnetic Scrolls wasn't a task that could be handled by most, so software houses started thinking of novel approaches to ultimately simplify gameplay and avoid frustration in players whose commands were not recognized not because they were trying to do anything wrong but simply because they were using different synonyms for the words the designer originally thought of.

To achieve this and work around the parser constraints, players had to input commands in a different way. A possible solution was found in offering a set of multiple choices: less engaging, maybe, but surely much easier to handle development wise. Sierra was one of the first companies experimenting with this approach, with simple games targeting children like "Troll's Tale" and "Dragon's Keep", both released in 1984, but the best example of this kind of games was probably Level 9 Computing's "The Secret Diary of Adrian Mole", based on the homonymous book series by Sue Townsend (Fig. 4.12).

Interestingly, a multiple choice approach wasn't used only for simple branching games but soon it became an effective way to link narrative and dialogues in games featuring more action oriented sequences, pioneering a type of gameplay that became commonplace later across different genres like action adventures and RPGs. A beautiful example of this early approach was "Law of the West" (Fig. 4.13).

The game followed the sheriff of a small town in the wild west across a rather eventful day where we could meet a varied set of characters. By using the joystick to select answers or, eventually, to grab our gun and aim around, the game managed to merge a more traditional adventure gameplay with the excitement of some well crafted shooting action sequences.

Experimenting with novel gameplay modes didn't stop here: the mid eighties were definitely a brainstorming time for adventure games, with designers constantly trying to find out new ways to retain the original engagement that was the trademark of purely text based adventures while making the whole experience more accessible and exciting.

Fig. 4.12 The Secret Diary of Adrian Mole (© 1985 Level 9 Computing): putting the player into the shoes of an almost 14 year old boy and all his little/big problems made for a very engaging game that perfectly matched the spirit of the original books

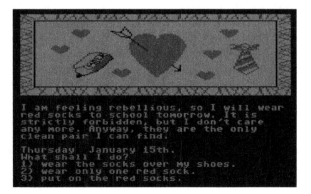

Fig. 4.13 Law of the West (© 1985 Accolade). How shall we deal with this guy? Our approach determines whether there will be a shoot out to end the confrontation or a more peaceful departure

Fig. 4.14 Lords of Midnight (© 1985 Beyond). While in the game, players could view the landscape from 8 different cardinal directions and move accordingly, examine their character state in detail and choose among other options which varied according to the specific situation (for example, recruiting new men or head to battle)

Inspired by Tolkien's epic novels, game designer Mike Singleton (1951–2012) successfully managed to merge adventure and strategy in "Lords of Midnight" (Fig. 4.14), a ZX Spectrum classic that was soon ported to the C64 as well. An epic adventure across 4,000 locations with 32,000 possible views where the player could control several independent characters in a quest to defeat the evil Doomdark. It was possible to accomplish victory in two completely different ways, adding significant replayability value to a package that was already impressive due to its scale and variety, with mountains, forests, towers, citadels and much more waiting to be discovered and visited.

Strategy, with the addition of tactical combat elements plus a well thought-out fencing action sequence, was also prominently featured in Sid Meier's first game to bore his name: "Pirates!" (Fig. 4.15). The game took place in an surprisingly vast and well researched Caribbean world set between 1560 and 1680 where, starting as a privateer for one of the ruling European empires, the player had to search for fame and fortune. Among the many original traits of the game, besides a constantly evolving world that made each game different and unique, it should also be remembered how the player couldn't actually die. His character instead gradually aged, ultimately becoming unfit for combat and forcing him into retirement. This

Fig. 4.15 Sid Meier's
Pirates (© 1987 MicroProse).
Originally developed on
the C64, "Pirates!" was
then ported to many other
computers and devices, from
1987 till this very day

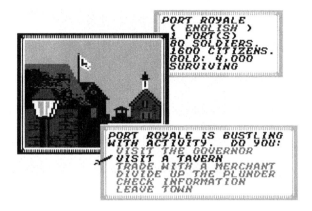

feature was cleverly used to rank players, who could finish the game by becoming as influential as a King's adviser or as poor as a beggar, but it was so ahead of its time that even the journalists at Zzap!64 couldn't really appreciate it and awarded the game an unimpressive 68 % score because of it.[2]

"Pirates!" wasn't the only groundbreaking game to be released in 1987 though. That year also saw another major evolution in adventure games: in the constant effort to simplify the genre while also leaving players plenty of options and freedom for manipulating objects like in the original text adventures, LucasFilm came up with a new engine to develop adventures following an innovative "point and click" approach. The first game using this method was "Maniac Mansion" (Fig. 4.16), which also gave its name to the engine itself: SCUMM, standing for "SCripting Utility for Maniac Mansion".

Playing the role of Dave, a young teenager whose girlfriend Sandy disappeared in Dr. Fred's sinister mansion up the hill, players had to investigate and rescue her together with a couple of school friends. While the "kidnapped girlfriend" setup was far from original, the game was crammed with detail and humour, with clever puzzles and over the top, amusing situations that were truly unique. Brief but entertaining cut scenes were seamlessly integrated with the game, simulating an effective camera work that brought the game closer to a cinematic experience. "Maniac Mansion" soon became a sensation and, today, is one of the most fondly remembered adventures of all time.

Point and click adventures were soon going to become a staple of LucasFilm and marked some of the most brilliant games the studio produced for 8 and 16 bit systems. Indeed, Maniac Mansion was soon followed in 1988 by "Zak McKraken and the Alien Mindbenders" (Fig. 4.17), which pushed the genre to even greater heights.

However, short listing a set of actions and clicking across interactive environment wasn't the only original possibility that game designers experimented with. Other games also started evolving this concept further by trying out icon driven Graphical User Interface (GUI) based systems to make adventure games

---

[2] Zzap!64 issue 29, p. 26.

Fig. 4.16 Maniac Mansion (© 1987 LucasFilm). Now objects could simply be selected on the screen by moving a cursor with the joystick and then operated according to specific commands already shortlisted in the GUI. Finally, language confusion or misunderstanding with the parser were things of the past!

Fig. 4.17 Zak McKraken and the Alien Mindbenders (© 1988 LucasFilm): a spiritual sequel to Maniac Mansion, Zak pushed the humour and the absurdity of some situations even further in a totally bizarre and eccentric mix that left no one indifferent

even more intuitive. Already in 1985 the computer game adaptation of the best-selling spy thriller "The Fourth Protocol" by Frederick Forsyth was crafted by Hutchinson Computer Publishing in a slick package where all relevant commands to track people and carry on investigations were simply accessible via icons. Activision, in 1986, tried to merge text adventures and clickable interfaces in the brilliant "Tass Time in Tone Town" (Fig. 4.18) but this approach was ultimately pushed to its limits a couple of years later to implement not only a few basic commands but a whole new icon driven language in Infogrames' "Captain Blood" (Fig. 4.19).

Developed by ERE Informatique, in "Captain Blood" players had to travel through a galaxy in search of five clones of themselves which had to be killed to recover life energy and survive. Searching the clones involved travelling from planet to planet and questioning different alien species for clues, but how to communicate with all these aliens when each spoke a different, incomprehensible language? This is where the innovative icon system came into play: to talk to them, and understand their replies, we had to rely on a visual, ideogram based language which gave a completely unique twist to the whole gameplay experience.

Fig. 4.18 Tass Time in Tone Town (© 1986 Activision): text plus GUI to immerse players in a cleverly allegorical and funny world where nothing can be taken at face value

Fig. 4.19 Captain Blood (© 1988 Infogrames). Whether we will get any useful information from this alien or not will depend on our communication skills with the provided icon/ideogram based language

4.3 Arcade Conversions

The early eighties were the golden age of arcades and coin-op machines so it is no wonder that, as soon as home computers started having enough power to somewhat replicate the main features of arcade games, adaptations started being released almost in sync with the appearance of new coin-op originals. Indeed, from the very beginning, the C64 proved itself to be a very good platform for trying to replicate arcade experiences at home due to its multimedia capabilities that could effectively approximate those of more powerful arcade machines.[3] Commodore worked closely with Bally Midway, licensing a few IPs that were then ported as launch titles or early releases for the C64.

One of these, and one of the first conversions that really managed to perfectly capture the spirit of the arcade original, was "Wizard of Wor" (Fig. 4.20).

Naturally, Commodore itself wasn't the only developer to work on arcade game ports for the new computer: third parties started closing licensing deals on their own and release very impressive products as well.

---

[3] Many arcade games in the early/mid eighties were customized multiprocessor machines based on CPUs from the Motorola 6800, MOS 6502 or Zilog Z80 families, i.e. they were mounting at least two CPUs that were working together to deliver a more impressive multimedia output.

Fig. 4.20 Wizard of Wor
(© 1981 Midway/1983
Commodore): a seminal
game in the maze and
monsters/aliens genre where
the lines between hunters and
prey get easily blurred

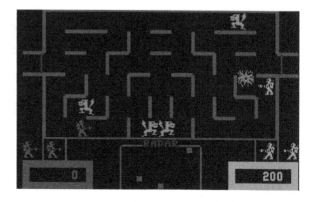

Fig. 4.21 Q*Bert (©
1982 Gottlieb/1983 Parker
Brothers). The colorful world
of Q*Bert was perfectly at
ease on the C64

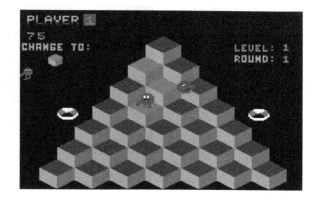

Parker Brothers, for example, released ports of classics like Q*Bert (Fig. 4.21) and Gyruss (Fig. 4.22) that were very faithful to the originals and are generally considered as the best home version ever released.

Konami, on the other hand, partnered with Imagine[4] to release home versions of several of their games, among which it is worth remembering classics like Yie Ar Kung Fu (Fig. 4.23), the Olympic simulation "Hyper Sports" (1985, showcasing also an impressive rendition of Vangelis' "Chariots of Fire" musical theme by Martin Galway) and the delightful school based action game Mikie (Fig. 4.24).

Capcom was another major player in the eighties arcade space that partnered with Elite for many of its C64 ports. "Ghost 'n Goblins" (Fig. 4.25) was a particularly successful example of their work but not all conversions were masterpieces: for example, classic arcade shooter Commando (1985) fell short of expectations and only a great soundtrack by Rob Hubbard made the game worth remembering.[5]

---

[4] In 1984 Liverpool based Imagine went bankrupt and its name was bought by Ocean Software. The latter kept using the Imagine brand as a sub-label for its games.

[5] An upgraded version of the game, very faithful to the arcade original, was released in 2014 and is available here: http://csdb.dk/release/?id=130973.

Fig. 4.22 Gyruss (©
1983 Konami/1984 Parker
Brothers). An original space
shooter that merged Galaga
with by Atari's Tempest,
Gyruss saw a hectic 360°
action in a desperate race
back from outer space to save
Earth

Fig. 4.23 Yie Ar Kung
Fu (© 1985 Konami/1985
Imagine). There would be no
"Street Fighter" or "Mortal
Kombat" games without "Yie
Ar Kung Fu": energy based
fighting games with fancy
moves start here

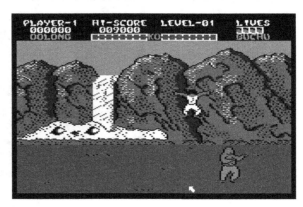

Fig. 4.24 Mikie (© 1984
Konami/1986 Imagine).
Can Mikie get all the hearts
around the classroom without
getting caught by the teacher?

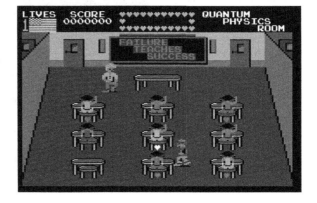

The following years proved the C64 could still produce quality conversions
despite arcade games getting bigger and bigger. Taito's Bubble Bobble (Fig. 4.26)
and Rainbow Island (Fig. 4.27) show how the C64 could effectively manage to

Fig. 4.25 Ghosts n Goblins (© 1985 Capcom/1986 Elite). Zombies and demons won't stop Sir Arthur in his quest to save Princess Prin Prin! Or maybe they will, considering this is one of the most difficult games ever made and the C64 version retains all the challenges of the original!

Fig. 4.26 Bubble Bobble (© 1986 Taito/1987 Firebird). Fast, hectic action for one or two players with many colorful sprites on screen at once made Bubble Bobble a favourite of many players that is still as fun to play today as it was back in the day

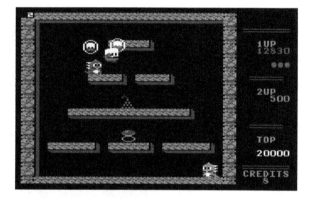

Fig. 4.27 Rainbow Island (© 1987 Taito/1990 Ocean). A great example of how to pack a 2 MB arcade game into a 64 Kb computer, while retaining all the charm and qualities of the original!

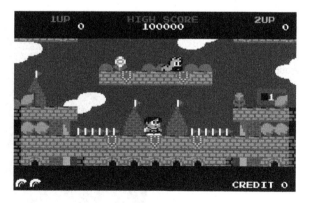

recreate this type of fast action games in detail. Even a few arcades from the end of the 1980s and early 1990s, like shoot 'em up platformer Toki (Fig. 4.28), managed to get competent ports before 8 bit systems ultimately had to abdicate in favour to 16 and 32 bit computers.

Fig. 4.28   Toki (© 1989 TAD
Corporation/1991 Ocean).
One of the last impressive
arcade conversions to grace
the C64 towards the end of its
lifecycle

4.4   Driving Games

Driving and racing games started appearing very early, with one of the first note-worthy example being Epyx seminal "Pit Stop" in 1983. Anyway, it was only with its sequel, "Pit Stop II" (Fig. 4.29) in 1984 that the genre really took off. Pit Stop II was a landmark not only because it emphasized head to head competition with a friend thanks to a very well designed split screen, but also for its variety of circuits and for adding a strategic element to the action, forcing players to determine the best time to get into the pit stop to refuel and change tyres: these, in fact, gradually wore out during the race and would ultimately blow up if neglected.

The popularity of racing games in the arcades supplied the C64 with a con-tinuous stream of ports. While some were just average, like the original "Pole Position" (Atarisoft, 1984) and Outrun (US Gold, 1988), others like "Buggy Boy" (Fig. 4.30) and "Turbo Outrun" (Fig. 4.31) did manage to transfer the arcade rac-ing experience very effectively.

Fig. 4.29   Pit Stop II (©
1984 Epyx). Different tracks
and an exciting split screen
view, allowing for engaging
head to head duels, granted
Pit Stop II a replayability
value that was unique among
racing games at the time

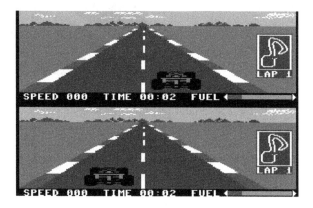

Fig. 4.30 Buggy Boy (©
1985 Taito/1987 Elite). A fun
and frantic racing experience
that scored a well deserved
97 % on Zzap!64 (issue 32,
p. 20)

Fig. 4.31 Turbo Outrun (©
1989 Sega/US Gold). Another
great arcade conversion that
also scored 97 % on Zzap!64
(issue 56, p. 8)!

Fig. 4.32 Revs (© 1986
Firebird). Designed by Geoff
Crammond, Revs was a
true Formula 3 simulation
based on the Ralt Toyota car.
The C64 version included
a detailed recreation of the
Silverstone Grand Prix and
Brands Hatch circuits

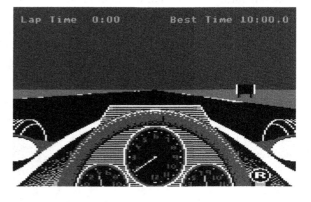

With the excitement of pure arcade style racers covered, original titles started
focusing more on driving from a simulation perspective, laying down the foun-
dation for modern racing games whose vision could only be fully accomplished
years later on much more powerful machines like Sony PlayStation.

Revs (Fig. 4.32) was the first game that successfully pushed the genre towards
a more realistic approach focused on physics and car handling. For the first time,

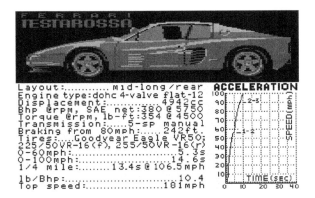

Fig. 4.33 Test Drive (© 1987 Accolade). One of the earliest attempts to recreate an accurate driving experience based on famous sport cars like the Ferrari Testarossa or the Lamborghini Countach

players could actually configure their car asset, for example by setting the angle of the front and back wings to determine the amount of downforce acting on the car during the race.

Qualifying laps and actual races on two accurately recreated tracks completed the most realistic driving experience available on home computers at the time.

In 1987 it was the turn of "Test Drive" by Accolade (Fig. 4.33), a franchise that remained popular for many years to come. Here four different car models were included for the player to take out for a trial on a relatively narrow street. If the player could manage to get till the end of the road safe and sound, he could get ranked accordingly and keep his very own Ferrari or Porsche, at least virtually.

The sequel, "Test Drive II: The Duel" (Fig. 4.34), arrived in 1989 and allowed players to experience two different cars, the Porsche 959 and the Ferrari F40,

Fig. 4.34 Test Drive II: The Duel (© 1989 Accolade). Chasing the computer car while being chased by the police made for some very exciting drives! Besides the main game, additional cars and scenery expansion disks were also released, pioneering a business model that was going to become very popular a few years later as soon as the Internet became a viable distribution method for downloadable content

Fig. 4.35 Lotus Esprit Turbo
Challenge (© 1990 Gremlin
Graphics). Lots of details
on the car here, coupled
with a fast and competitive
experience for one or two
players alike, made this game
a success

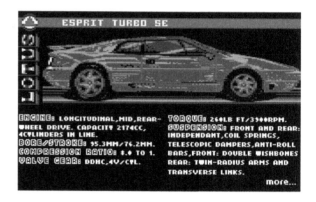

across two game modes: racing against the clock or against a computer opponent. Both modes took place across a trafficked highway, with police patrols ready to chase after the driver as soon as the speed limit was broken.

Still in 1989 Accolade also released another excellent driving game, "Grand Prix Circuit". This time the game left the countryside for a proper Formula 1 racing experience where players could choose between different settings that altered the realism of the game and car handling. Ranging from a pure arcade experience, where the engine couldn't blow up and gears were changed automatically, to more realistic conditions. In this way, the game successfully appealed to both arcade fans as well as those who liked a more true-to-life driving simulation.

Anyway, Accolade wasn't the only company striving for excellence in the genre. A last, comprehensive attempt to synthesize all the different approaches that made older games memorable was tried by Gremlin Graphics in "Lotus Turbo Challenge" (Fig. 4.35). Released in 1990, Lotus was notable for adding some simulation elements, modelling the player's car after the Lotus Esprit Turbo SE, into an "Out Run" kind of set up with a challenging run across 32 tracks, besides a split screen for 2 players action like in "Pit Stop 2".

4.5   Edutainment

In the early days, when the market was still virgin and it was not so clear which type of software was in actual demand, it was important to attract a new user base who had no previous experience with computers and technology. Back then, using computers for educational purposes seemed like a very natural choice to show parents how even simple games could help in making their kids smarter by teaching basic math or reading skills. Different companies, like Fisher Price and Spinnaker, started making a name for themselves in the new educational niche via several titles that were generally met with moderate success. Commodore itself was among the most active publishers of early educational software, releasing titles to learn Braille or rough but picturesquely named little utilities and games like "Big

Fig. 4.36 Visible Solar System (© 1982 Commodore). Despite the very low resolution graphics, many kids for the first time had the illusion of exploring the main planets and moons of our solar system. While not strictly a game, this was one of the earliest commercial simulations developed with educational purposes on home computers

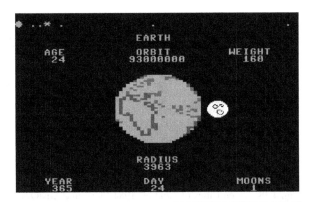

Fig. 4.37 Tooth Invaders (© 1982 Commodore). A simple game to make kids understand the importance of a proper oral hygiene, "Tooth Invaders" can be considered a pioneer in the serious games and games for health fields

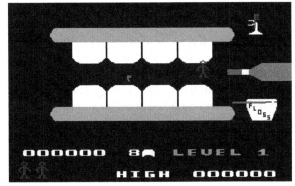

Ohm's Law", "Bomb Addition" or "Brain Crane". Among this first generation of software, two titles in particular stand out: "Visible Solar System" (Fig. 4.36) and "Tooth Invaders" (Fig. 4.37).

However, due to the rapidly increasing quality of software and games being released, simple products like "Tooth Invaders" and "Visible Solar System" aged pretty fast. It didn't take long for kids in the early eighties to understand what was possible on their C64 and start dreaming big, demanding better and more engaging software especially if this had some sort of educational aspect with no aliens to blast left and right.

The first company that understood this perfectly and not only pushed the concept of "edutainment" to the next level but actually defined the word in the first place, was Electronic Arts via Trip Hawkins, its founder. He was, in fact, the very first person to adopt the term while presenting EA's latest game "Seven Cities of Gold" (Fig. 4.38) to the press in 1984, stressing the high level of attention that went into recreating a particular historical period, the early colonial era in this case, thereby making the game also a very effective tool to learn about history in a natural and fun way.

In "Seven Cities of Gold", developed, among others, by legendary game designer Dan Bunten,[6] the player took the role of a Spanish conquistador set to explore the New World. Lost cities and hidden treasures were awaiting and both

---

[6] Dan Bunten (1949–1998 as Danielle Bunten Berry).

Fig. 4.38 Seven Cities of Gold (© 1984 Electronic Arts/Ozark Softscape). Ready to embark in new, uncharted but historically accurate adventures

raiding native villages or a more peaceful trading approach were viable options in a vast and rich world that had no equals in earlier games. There were no pre-defined set routes in place here, leaving the player completely free to choose his strategy and, ultimately, his destiny. The significance of "Seven Cities" doesn't stop at its game design innovations, though, but also stands out from a technical perspective: for example, for the first time, maps were streamed from the disk while playing, allowing for a smooth and streamlined gameplay that didn't need to be interrupted by annoying loading times whenever the player moved to a different region of the in-game world.

The game became an instant classic and, across all its different ports, sold more than 150,000 units, a really exceptional result for an industry in still its infancy like computer games in the mid eighties.

The same concept was then polished even further in the sequel "Heart of Africa" in 1985 (Fig. 4.39): this time the action was set in the late XIX century Africa and the game brought the newly born genre of "edutainment", here once again focused on history education via realistic adventures and settings, already to maturity.

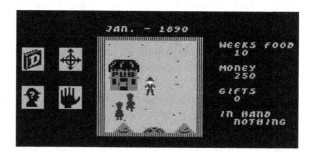

Fig. 4.39 Heart of Africa (© 1985 Electronic Arts/Ozark Softscape). Learning by exploring the last hidden areas of the world 400 years after the events depicted in "Seven Cities of Gold". The game was still a very good commercial success for EA, selling about 50,000 units, but wasn't able to match the same success and cult status that the earlier title had

Fig. 4.40 Ghostbusters (©
1984 Activision). Trying to
capture a ghost by framing
it within two proton streams
to suck it into a trap wasn't
always an easy task!

CITY'S PK ENERGY: 701 $1000

4.6 Movie Tie-ins

After the colossal E.T. fiasco by Atari and Warner in 1982, with 25 million dollars paid to Steven Spielberg to license the movie rights for its 2,600 game adaptation,[7] the gaming industry definitely had to be less bold and rework the ways it did business with related entertainment industries. However, the eighties were a time of movie blockbusters, especially in the sci-fi and action genres where new technologies could effectively be used for groundbreaking special effects, so it soon became obvious that, if done right, sharing IPs across media was a natural and potentially very successful move. Movie tie-in became increasingly popular and players soon began expecting games where they could re-live in an interactive form the heroic actions seen on the silver screen, now on their own TVs thanks to their computers.

One of the first releases that showed how well games and movies could work together, was Activision's Ghostbusters (Fig. 4.40), based on the 1984 hit. Designed by David Crane and with an excellent musical rendition of the movie themes by Adam Bellin, the game managed to capture the spirit of the movie pretty well, alternating frantic action sequences around New York with strategic pauses to buy and upgrade ghost chasing equipment, before the final showdown with the Marshmallow Man and Zuul.

Unfortunately, not all early conversions were met favourably by the press and fans alike. For example, despite trying to follow the movie plot closely, the 1986 game developed by Electric Dreams and based on the well known "Back to the Future" movie featuring Michael J. Fox, fell short of expectations and gamers had to wait till 1987 for two more excellent games: Aliens (Fig. 4.41) and, especially, Platoon (Fig. 4.42).

The first, based on 1986 James Cameron's classic and sequel to the 1979 "Alien" movie by Ridley Scott, let the player coordinate the work of six space marine sent to investigate a space colony across 200 locations with the ultimate aim of destroying the Alien Queen and her eggs before escaping. The game

---

[7] See Dillon:"The Golden Age of Video Games", CRC Press, 2011, p. 73.

Fig. 4.41 Aliens (© 1987
Electric Dreams). A first
person perspective with a
HUD (heads-up display).
Outlining also biological
data, showing the marine's
heartbeat rate, contributed
significantly in enhancing the
whole atmosphere and sense
of imminent danger

Fig. 4.42 Platoon (©
1987 Ocean). Fighting in
narrow quarters across an
underground first person
scene of the game. The
claustrophobic tunnel 3D
view was one of the high
points of the game

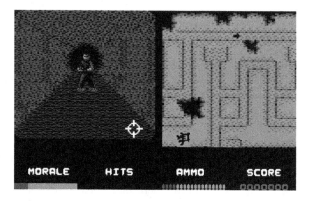

successfully managed to build an exciting and scary atmosphere and was very well
received by the specialized press.

It was Platoon, though, that significantly raised the quality bar. Thanks to different movie inspired scenes, set across a variety of environments ranging from
jungles to villages and tunnels, gameplay was rich and varied. Besides, changes
in camera perspectives, alternating standard 2D side views with effective first and
third person actions, also added excitement and a very cinematic feeling.

The only part where the game possibly fell short was in depicting the ugly side
of war without any form of glorification, which was the original aim of the 1986
movie by Oliver Stone.[8] Platoon probably tried to achieve this by being extremely
difficult and brutal, besides penalizing players if they killed unarmed villagers.
Yet, its high difficulty level most likely provoked more frustration in players than
thoughtful reflections on the morality of war.

Anyway, when talking about action movies in the eighties, two names in particular come to mind: Sylvester Stallone and Arnold Schwarzenegger. Indeed,
many of their action flicks would have been suitable for videogame tie-ins and,
unsurprisingly, they were.

---

[8] For having such a game we will have to wait a few more years till "The Lost Patrol" (Ocean)
on the Commodore Amiga in 1990.

Fig. 4.43 Rambo: First
Blood part II (© 1986,
Ocean). Arguably the best
Stallone based game on the
C64

The games based on Stallone's movies, though, didn't manage to impress the specialized press.

"Cobra", released in 1986 by Ocean, scored a shameful 13 % on Zzap!64 and only the vertical scrolling shoot 'em up "Rambo: First Blood Part II" (Fig. 4.43) managed to achieve a decent score.[9] Its sequel, "Rambo III" (Ocean 1988) again was scored at only 47 % by Zzap!64 journalists.

Unfortunately, also Schwarzenegger beginnings as a video game hero were not much more encouraging: the scrolling beat 'em up "The Running Man" (1989, Grandslam) scored a low 44 %[10] but, luckily, the following attempts fared much better. "Total Recall" (Fig. 4.44) obtained a much more honorable 76 %[11] and "Terminator 2" (Fig. 4.45) surged to very high levels with a 89 % review score[12] thanks to its well designed graphics, music and action that properly followed the original plot.

Among the several other tie-ins, a few more need to be remembered. 1989 was actually a very good year with several interesting releases: Batman finally moved from comics to the silver screen and then started becoming more prominent as a video game character as well. The game licensed from Tim Burton's 1988 movie was published by Ocean (Fig. 4.46) and was heralded as the finest superhero game to date.

But there was much more than Batman: "The Untouchables" (Fig. 4.47) based on the 1987 crime drama directed by Brian De Palma was another "Gold Medal" on the issue 55 of Zzap!64. The press highly prized the game's variety, with six different stages that provided exciting gameplay that never got boring thanks to top-notch action presented from different perspectives, following the same approach that was so successful in Platoon.

Ocean cemented its excellent reputation even further with the tie-in for the 1987 movie RoboCop directed by Paul Verhoeven, released at the beginning of

---

[9] 65 %, Zzap!64 issue 10, p. 23.

[10] Zzap!64 issue 53, p. 70.

[11] Zzap!64 issue 71, p. 76.

[12] Zzap!64 issue 78, p. 14.

Fig. 4.44 Total Recall (©
1991, Ocean). Unfortunately
the intense stare of Arnie
didn't help this mix of
combat, explorations and
platforms-and-ladders to
achieve everlasting success

Fig. 4.45 Terminator 2 (©
1991, Ocean). A fight with a
cop opens the game, which
was divided into different,
independent sections (or
"mini-games", as modern
gamers would likely call
them today) connected by
simple cut scenes to help
players relate the action to
the original movie

Fig. 4.46 Batman: the
Movie (© 1989, Ocean).
Great graphics and sound,
together with a well
calibrated mix of platforming,
driving across Gotham City
and puzzle levels, impressed
critics and players alike,
helping the game to score
96 % on Zzap!64 (issue 55,
p. 9)

1989. The game (Fig. 4.48) offered a compelling side scrolling shooting action
that, while not particularly original, was enough to satisfy the fans of the film.

Sequels soon followed, both in theatres and on C64 screens, with Robocop 2
being released in 1990 and Robocop 3 in 1992. The former emphasized platforming action with a more cartoonish graphical style while the latter also added a few

Fig. 4.47 The Untouchables (© 1989, Ocean). The game featured also jazzy musical tunes typical of the time, effectively used to depict the wild yet carefree Prohibitionist years in Chicago

Fig. 4.48 Robocop (© 1989, Ocean). The beginning of the game: it will be a long and tough journey to defeat OCP and restore control in Detroit

levels played from a first person perspective where players had to aim a crosshair to shoot enemies across a scrolling background. Overall they didn't add anything particularly original to the mix but, oddly, these, together with Navy Seals, another tie-in conversion by Ocean of a 1990 movie with Charlie Sheen and Michael Biehn, were among the rare cases where the games' critical receptions were actually much higher than those of the respective movies: Robocop 2 obtained a 90 % on Zzap!64[13] and Robocop 3, like Navy Seals, scored even higher, getting a 92 %.[14] On the other hand, the movies were badly criticized by the specialized press, as reflected by their average critics score of only 4.5 and 3.1 out of 10 respectively for the Robocop movies and 4.2 for Navy Seals, as reported by Rotten Tomatoes website.[15]

---

[13] Issue 69, p. 8.

[14] Robocop 3 was reviewed in issue 82, p. 16. Navy Seals was in issue 69, p. 18.

[15] See http://www.rottentomatoes.com/m/robocop_2/, http://www.rottentomatoes.com/m/robocop_3/ and http://www.rottentomatoes.com/m/1029427-navy_seals/.

4.7 Platformers

Platformers were born in the arcades in 1980 with "Space Panic" by Universal and grew up quickly, reaching maturity already 1 year later thanks to Nintendo's "Donkey Kong". Then, after the NES established itself as the leading gaming console worldwide thanks to games like Super Mario Bros (Nintendo 1985), most people started considering this genre as an exclusive playground for dedicated systems across the 8 and 16 bit era. Nonetheless, computers like the C64 could still offer some truly original experiences that shouldn't be forgotten. As we will see, early home computer platforming games were often breaking new grounds by effectively merging elements from different genres, resulting in unique and innovative game concepts that were not available anywhere else.

Exploration of vast areas was a common and pervasive theme in this type of games, starting with Broderbund's "Lode Runner" (Fig. 4.49) in 1983. Here players were put in front of 150 challenging levels to collect gold while avoiding robotic sentinels. If the available mazes were not enough, a level editor was also included (a first for the industry), allowing players to become level designers and build their own levels to continue the fun.

Off to a strong start, the following years were even more remarkable for computer platformers: Epyx released "Impossible Mission" (Fig. 4.50), an outstanding game not only from a mere technical perspective (running and somersaulting animations were smooth and of cinematic quality while audio was simply astonishing with sampled speech which included the opening line "Another visitor! Stay awhile, stay forever!" or the calling "Destroy him, my robots!" that still echo in many players' ears to this very day) but also for merging pure jumping platforming action to puzzle elements and a 007-like story, with the player infiltrating the secret hideout of an evil genius, aptly named Prof. Atombender, to vanquish his world domination plans.

Still in 1984, Datasoft managed to effectively synthesize platforming with fighting in "Bruce Lee" (Fig. 4.51). While the legendary martial artist was recognizable only in the splash screen, this reference was enough to draw young players into a new oriental adventure. The simple but satisfying fighting action against a black

Fig. 4.49 Lode Runner (© Broderbund, 1983). While jumping was not an option here, players could set up traps to temporarily halt their pursuers by opening holes on the floor. This was similar to the original "Space Panic" arcade game but here it was implemented differently to allow for a much faster and more exciting gameplay

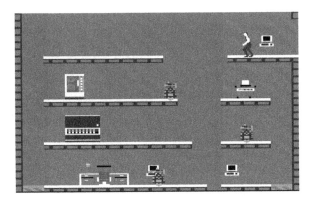

Fig. 4.50 Impossible Mission (© Epyx, 1984). Defeating Prof. Atombender in his hideout not only made for a great challenge but also made Impossible Mission the first story based platform game that actually had an ending

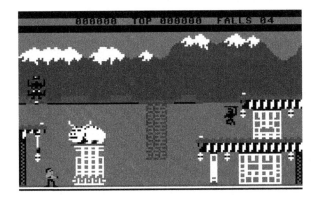

Fig. 4.51 Bruce Lee (© Datasoft 1984). The black ninja enters the opening screen pointing straight to our hero. Green Yamo, the sumo wrestler, will soon follow resulting in hectic chases with punches and kicks being thrown all over the screen. In a very interesting and original twist of the game, the two enemies could actually hit each other and be lured into traps, resulting in some very rewarding moments where the player could outsmart his opponents by effectively using the environment

ninja and a green sumo wrestler complemented very well the quest for collecting lanterns, opening secret passages to progress into a fortress full of environmental hazards and ultimately defeat an evil wizard.

A more cartoonish approach was instead used in "Monty on the Run" by Gremlin Graphics (Fig. 4.52). Here, accompanied by a spectacular soundtrack courtesy of Rob Hubbard, players had to help Monty the Mole in escaping across a set of devilish screens after being jailed for stealing coal in a previous game.[16] Classic platforming action (with Monty not just jumping but somersaulting like

---

[16] "WANTED: Monty Mole" (Gremlin Graphics, 1984) inspired by the miner's strike in the early 1980s in the UK.

Fig. 4.52 Monty on the Run
(© Gremlin Graphics 1985).
Can we help Monty flee the
country?

the secret agent in "Impossible Mission"!) was mixed with numerous environmental puzzles that required different items to be solved, including some that had to be chosen at the beginning of game from a "freedom kit". If chosen wrongly, the player would soon or later get stuck somewhere in the middle of the game and realize he had to start all over again with a different combination of tools.

Despite their charm, platform games like "Monty on the Run" could actually be quite frustrating due to their high difficulty and to the (likely) possibility of being unable to finish the game if any of the required items was missed, forcing the player to begin a new game and go through all the levels once again. To make the experience a little more forgiving, Big Five Software in their 1985 game "Bounty Bob" (Fig. 4.53) started hiding some "cheat codes" around the game world, providing players a way to skip early levels and not being forced to replay the game every time all the way from the very beginning.

1985 was also the year Nintendo released "Super Mario Bros" on the NES, providing a new role model all platform games had to be measured against from then on. Super Mario's influence, featuring colorful worlds and secret areas, very smooth scrolling and perfectly tuned animations and challenges, would be apparent also on the computer scene where several games tried to replicate its qualities. The most infamous example was "Great Giana Sisters" by Rainbow Arts (Fig. 4.54)

Fig. 4.53 Bounty Bob (©
Big Five Software 1985).
Two years in the works,
Bounty Bob was a very
complex and well thought
out sequel to Miner 2049'er
that added several original
and unique elements to its
predecessor

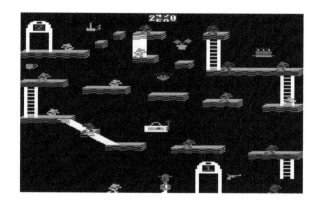

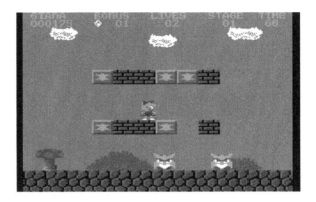

Fig. 4.54 Great Giana Sisters (© Rainbow Arts 1987). "Look, she has a sister, not a brother and they are great, not super!" unfortunately these arguments were not enough to convince Nintendo's lawyers about Giana's legitimacy as a videogame heroine

that, unfortunately, referenced Super Mario a bit too literally. A very high quality production in its own right, it also triggered a lawsuit from Nintendo that forced the game to be soon pulled off the shelves and be withdrawn from the market.

Regardless of Giana's legal misfortunes, the focus on exploration and attention to detail that made Super Mario such a great game did set a new standard of quality that was apparent also in several other, truly original, games. "Myth: History in the Making" (Fig. 4.55) was one of such games. Here fantastic graphics supported a multidirectional platforming action that often broke the side scrolling template established by Super Mario. Players were left breathless across different levels where platforms and ladders gameplay merged perfectly with action adventure elements in a quest across time that featured tough combat and non trivial puzzles.

Similar approaches were tried in other games that are still remembered as classics, like "Rick Dangerous" (Firebird 1989), where levels were tied together by an Indiana Jones-like adventure narrative or the extremely polished "Creatures" (Thalamus 1990), a very colorful and fantasy themed game where more traditional scrolling sections were interleaved with single screen "torture" levels that required both thinking and fast reflexes, resulting in an effective change of pace.

Fighting action mixed to platforming gameplay was also central to "First Samurai" (Fig. 4.56) that, together with "Mayhem in Monsterland" (Fig. 4.57),

Fig. 4.55 Myth: history in the Making (© System 3 1989). The first level was set in ancient Greece. Get ready to fight several monsters and beasts from Greek mythology here, including Cerberus, Medusa and the Hydra!

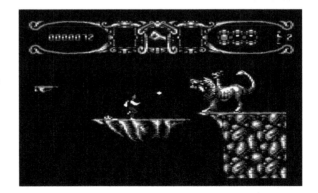

Fig. 4.56 First Samurai (© Image Works 1992). An engaging quest to avenge his master's death will take this lonely samurai across time and space in a fantastic world to ultimately face the Demon King

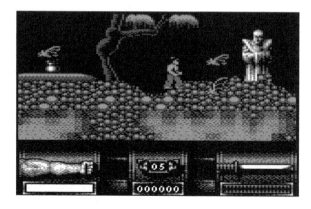

Fig. 4.57 Mayhem in Monsterland (© Apex Computer Production 1993). A stunning production that was on par with games developed on much more powerful 16 bit machines, Mayhem was awarded a perfect 100 % score on Commodore Format (11/93) and is generally considered as the swan song of the C64

was one of the last great releases on the C64. These last productions showcased, once again, the terrific technical achievements that were accomplished in 10 years of hard, passionate work on the same machine and would have been considered impossible when the computer was first released.

4.8 Puzzles

Arguably, the most famous puzzle game of all time is Tetris which, by selling 8 and a staggering 33 million units when released on Nintendo's NES and Game Boy respectively in 1989, took the world by storm and established the genre in the public consciousness once and for all. Still, Tetris wasn't a new game by then: it was first developed in 1984 by Alexey Pajitnov, a computer engineer working at the Dorodnitsyn Computing Centre of the Soviet Academy of Sciences in Moscow, and then started being licensed to many different platforms. The Commodore 64 version (Fig. 4.58), released in 1988, was actually one of the very best ports that, besides a smooth gameplay, could also offer a very engaging soundtrack to enhance the overall experience.

Fig. 4.58 Tetris (©
Mirrorsoft 1988). The C64
version of this timeless
classic showed how an
original 26 min soundtrack
(by Wally Beben) could
push the already engaging
gameplay of matching
and aligning the different
geometrical shapes to new
heights

Fig. 4.59 Sokoban (©
Spectrum Holobyte 1988).
This was a very addictive
transport type of puzzle that
required players to plan their
moves very carefully, almost
like a game of chess

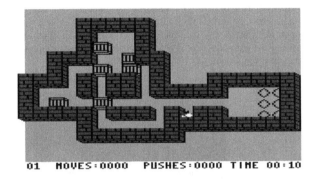

Anyway, puzzle computer games didn't start with Tetris. In 1981 in Japan, for
example, Hiroyuki Imabayashi designed a game called Sokoban (Fig. 4.59) where
players took the role of a janitor trying to move big crates around a warehouse and
match specific configurations. The game could quickly reach a high, often frustrating, complexity[17] and was published with good success in Japan in 1982. Western
audiences, on the other hand, had to wait till 1988 when Spectrum Holobyte
finally released it for the C64, MS-DOS and Apple II computers where it quickly
grow in reputation and popularity spawning many imitators.

The first game destined to be a classic and to be originally designed on the popular
8 bit computers of the 1980s was "Boulder Dash" (Fig. 4.60) which appeared in 1984.
Surprisingly, it not only had several sequels but was also the first home computer title,
together with Broderbund's side scrolling shooter Choplifter, to be converted into an
arcade game in 1985, taking the inverse route than most other licensed games did.

Besides these evergreens, there were many other games released during these
years that, even though they didn't quite reach the same level of popularity, were
actually hidden gems that shouldn't be forgotten. Activison's "Zenji" (Fig. 4.61)
and Mirrorsoft's "Bombuzal" (Fig. 4.62) definitely fall into this group.

---

[17] Sokoban can actually be analyzed via the theory of computational complexity and found
applications in universities where it was used to research and model robotic AI to actually move
around different warehouse settings.

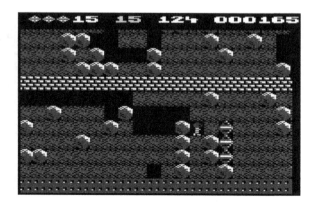

Fig. 4.60 Boulder Dash (© First Star Software 1984). Playing as a little guy named Rockford, players had the task of collecting diamonds while digging underground. Tunnels would not only open Rockford's way but could also set into motion massive boulders that would fall, possibly killing the cute character. Careful planning and fast running were definitely needed here when deciding where to dig

Fig. 4.61 Zenji (© Activision 1984). Here players act as a rotating face trying to connect all the different parts of a maze by rotating the section we are on (similar gameplay to Zenji can be found in much later games like the intruding and hacking minigame in the AAA blockbuster title Watchdogs (Ubisoft 2014)). Bonus points to collect and enemies to avoid complete a simple but challenging design

Bombuzal was particularly noteworthy due to its high polish and impressive graphics that featured a cute cartoonish creature with a dangerous task: clearing all levels of bombs and mines by making them explode without getting trapped. A handy password system was included to allow players to restart from any previously unlocked level and, in a unique kind of collaboration, a few of the over one hundred levels were laid out by some among the most well known game designers of the time, including Jon Ritman, Geoff Crammond and Jeff Minter.[18]

---

[18] Amusingly, but not surprisingly, Jeff's level exploded to reveal the shape of a lama.

Fig. 4.62 Bombuzal (© Mirrorsoft 1988). The loading screen for the game, which featured both an impressive isometric view as well as a more traditional 2D overhead perspective to help planning in the larger and more complex levels

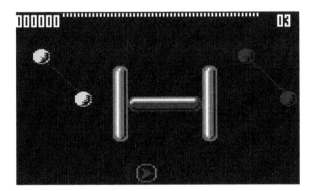

Fig. 4.63 E-motion (© US Gold 1990). The atoms, represented by spheres and often bound together into molecules, moved according to Newtonian physics with their bindings stretching and contracting accordingly. The player controlled a miniaturized ship that could set the atoms into motion by pushing them around. Barriers placed around the 50 available levels added challenge to the already tricky task

With the new decade starting, game designers also found inspiration from a variety of different subjects, including mirrors and light beams in Deflektor (Gremlin Graphics, 1990) or even chemistry, as shown by two very interesting games, E-Motion (Fig. 4.63) and Atomix (Fig. 4.64). In the former we had to destroy atoms floating in a microscopic world by making them collide with each other. Collision between atoms of different colors would give different results, making for an often chaotic yet fascinating gameplay that could be experienced also in an engaging two players co-op mode. The latter instead asked players to form actual molecules by combining atoms within a strict time limit across a playfield reminiscent of a sokoban game.

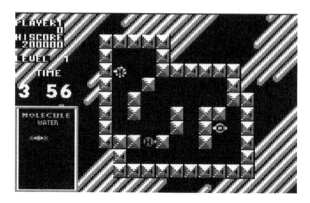

Fig. 4.64 Atomix (© Thalion 1990). The first out of thirty levels asks us to form a molecule of water by bringing together two atoms of hydrogen and one of oxygen

Fig. 4.65 Swap (© Palace Software 1991). A game with "candy attractive main graphics" as stated by Zzap!64 (issue 78, p. 61). Notice the small board on the right: players could take those supplementary tiles and place them in the main grid to refill empty spaces and clear isolated tiles if needed. "Avalanches" could also be triggered to make tiles fall down to fill existing holes

Curiously, both games had an image of Einstein in their splash screen, likely to emphasize the intelligence needed to accomplish the required tasks.

Among the puzzle games released during the later part of the C64 lifespan, at least one more deserves to be mentioned: Swap (Fig. 4.65). This can actually be considered a sort of granddaddy to all tile swapping based games that became overly popular with casual players on both web browsers and smartphones in the last few years. However, unlike most modern games, here tiles were not swapped to achieve a "match-3" combination: just having two tiles of the same color next to each other was enough to destroy them and players could also swap tiles that didn't result in any elimination. Different shapes (squares, hexagons and triangles) kept the game engaging and varied across its many levels.

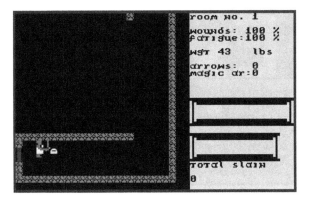

Fig. 4.66 Dunjonquest: Temple of Apshai (© Strategic Simulations/Epyx 1979/1983). The beginning of the game, after having defined and equipped our character. While the graphics were obviously lacking in detail, the manual provided evocative descriptions that had to be read along when entering each room for a proper understanding of the game

4.9 Role Playing Games

Few genres have always received as much attention and love throughout every hardware generation as Role Playing Games (RPG). Having their roots in the Dungeon & Dragons tabletop games invented by Gary Gigax and Dave Arneson in 1974, the popularity of computer RPGs can be traced back to 1979 when Automated Simulations (soon to be renamed Epyx) released "Dunjonquest: Temple of Apshai" on the TRS-80. It was an instant success and the game was then ported to other systems, with the C64 version being released in 1983 (Fig. 4.66).

Computer Role Playing Games became even more popular in 1981 when the Ultima and Wizardry series were released on the Apple II. Unfortunately, these classics reached the C64 with some delay: Ultima II and Ultima III (Fig. 4.67), originally published in 1982 and 1983 respectively, where ported in 1984 while the original Ultima (Fig. 4.68) was published only in 1986. Wizardry (Fig. 4.69) suffered an even longer delay, with the first three chapters of the series being published all together as late as 1987.

Overall, the first 6 episodes of the Ultima series (with Ultima VI being published in 1990) and up to Wizardry V (1988)[19] were released on the Commodore 64. Luckily, these delays weren't critical for all C64 RPG fans and several other great games appeared in the meantime: in 1985 Electronic Arts published "Tales of the Unknown: The Bard's Tale" (Fig. 4.70) that was soon to be known simply as "The Bard's Tale" and was going to set a new standard in RPG games by offering updated visuals and a somewhat more streamlined and user friendly gameplay.

---

[19] With the exception of Wizardry IV which was not released on the C64.

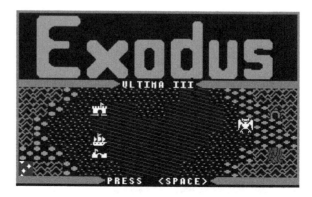

Fig. 4.67 Ultima III: Exodus (© Origin Systems 1983/1984). For the first time in the Ultima series, players could now create and command a party of characters in their quest to defeat the mysterious Exodus and also had an atmospheric soundtrack to accompany them in their quest. Set in the vast land of Sosaria, like the earlier games, adventurers could travel by land or sea, enter towns with plenty of people to interact with and, naturally, explore complex dungeons to fight vicious monsters

Fig. 4.68 Ultima I: The First Age of Darkness (© Origin Systems 1981/1986): the first chapter of the saga designed by Richard Garriott, aka Lord British, appeared dated when released on the C64. Nevertheless the depth of its scenario, with the player summoned to defeat the evil wizard Mondain, made up for the simplistic graphics and can still absorb any young gamer into its fantastic world full of adventures and surprises

Set in the city of Skara Brae, in a fictional medieval world hinting at Celtic and Nordic mythology, players began at the Adventurer's Guild to assemble a party made of mages, warriors, rogues, bards etc., each of them having different specialties and characteristics. Then they could freely wander around the city and enter not only weapon shops, taverns and healing churches but also any building on the map. Getting around, and especially exploring towers and underground dungeons, would likely challenge the player's party with many types of monsters in a quest to become powerful enough to finally face the nefarious wizard who turned the once cheerful city into a dangerous and dark place.

Fig. 4.69 Wizardry: proving Grounds of the Mad Overlord (© Sir-Tech 1981/1987). Developed by Andrew Greenberg and Robert Woodhead, the game allowed players to create a complex party of up to six characters (which had to be imported into the second and third instalments to keep playing) and then set for a type of adventure typical of the "dungeon crawling" genre, i.e. exploring a multi-level dungeon to gain experience by defeating monsters and find rare items to keep improving the different characters. The game is often remembered for having an unforgiving level of difficulty and could take hundreds of hours to finish

Fig. 4.70 Tales of the Unknown: The Bard's Tale (© EA 1985). Enemy portraits were animated and depicted effectively while it was also easy to give orders and understand how the fight evolved thanks to clear comments. With a huge variety in both opponents (128 different types of monsters were inhabiting Skara Brae) and party abilities, engaging combats with plenty of options were always guaranteed

The Bard's Tale saga continued very successfully in two sequels: "Bard's Tale II: the Destiny Knight" (released in 1986) and the "Bard's Tale III: the Thief of Fate" (1988), which also added a starter dungeon to help beginners and featured even more spells and monsters (500!) than ever before, besides more refined graphics and sound effects.

Another trilogy that is fondly remembered by many C64 RPG connoisseurs is "Phantasie" by Strategic Simulations Inc. The first chapter[20] (Fig. 4.71) was

---

[20] Phantasie II and III were released in 1986 and 1987 respectively.

Fig. 4.71 Phantasie (© SSI 1985): the flexible combat system, allowing each character to choose among several different choices and plan for an overall team strategy, was one of the high points of the game. Here the party is facing a big group of orcs. Will they survive?

released in 1985 and put players in the shoes of an adventurer seeking fame and fortune across the vast Isle of Gelnor, a land terrorized by the sorcerer Nikademus and his Black Knights.

The game proceeded in a traditional way, with a character generation and party formation system before being able to explore the land and its secrets, but was original in how it let players split experience and gold among party members. It also allowed for interesting combat strategies while presenting dungeons as an unfolding 2D map with a basic fog of war effect instead of trying a 3D wireframe approach like many other games.

Not all games came in trilogies, though. For example, Interplay's post-apocalyptic masterpiece "Wasteland" (Fig. 4.72) had to wait till 2014 for a proper sequel despite being considered one of the most fascinating games in the genre and being a clear inspiration for later franchises like "Fallout" (Interplay, 1997). The originality of the design was apparent from the very beginning, with a character creation system that allowed for complete freedom without forcing players to specialize into a predefined "class": any of the four characters forming the player's party could in fact learn, in principle, any skill and use any equipment since their specialization could be decided later as the game progressed. The open world setting was also very impressive: a vast land where every town and settlement looked

Fig. 4.72 Wasteland (© Interplay/Electronic Arts 1988): the main quest involved finding and destroying a factory producing killing robots, but players could simply decide to take other missions and explore the rich world if they so decided

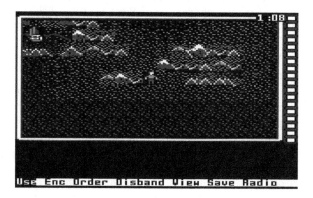

Fig. 4.73 Legend of
Blacksilver (© 1988 Epyx).
Big and colorful graphics
together with an easily
accessible control menu made
Blacksilver stand out and
please both newcomers and
experienced players

unique and where players could simply lose themselves in countless side quests, even ignoring the main storyline. Combat was easy to understand and most situations in the game could be faced in a multitude of ways, allowing for a truly unique and engaging experience. Last but not least, to enhance its immersive qualities even further, the game also included a book of paragraphs that provided more information about the world and specific locations, reminding the approach of another pioneering efforts, "Temple of Apshai".

At the same time, Epyx also continued its long standing tradition of excellent RPG productions with "Legend of Blacksilver" (Fig. 4.73), a game that, while not particularly original and quite linear in structure, did put lots of efforts in providing beautiful and bold graphics to make its world come to life.

One of the last great RPG to appear on the C64 was Ubisoft's (at the time named "Ubi Soft") B.A.T. in 1990 (Fig. 4.74). Set in a futuristic universe, here the player took the role of an agent from an ultra secret organization named "Bureau of Astral Troubleshooters" (hence the game name) with the task of stopping a group of madmen just escaped from prison and threatening to kill everyone in the city of Terrapolis, the only metropolis in the otherwise barren planet of Selenia, unless given ownership of the city itself.

Fig. 4.74 B.A.T. (© Ubi
Soft 1990). The engaging
cyber punk atmosphere was
a nice change to the usual
fantasy and dungeons settings
and players could customize
their characters by increasing
skills like "psychology" or
"evaluate" to improve in their
people skills and ability in
assessing specific situations

B.A.T. was an extremely ambitious game, not only for its control system, which was completely joystick driven by using icons and menus for command choices, but especially for trying to give players total freedom within a truly living environment with countless (more than 1,100 to be precise) different locations to explore, including parks, discos, restaurants, hotels, shops etc. Players here had the opportunity to interact with any NPC that happened to be around: anyone, in fact, could be stopped on a street to start a conversation, possibly uncovering some useful clues.

If these features were not enough, the game showcased also action sequences, including a flight simulator, and included a programmable computer (the "Bidirectional Organic Bioputer", or B.O.B., which was implanted in the agent's arm for easy access) that could be used not only for reviewing the player's stats at any time but also for actually programming some simple tasks, for example issuing warnings in specific situations, like when risking dehydration, or triggering an adrenaline rush if pursued by enemies. All this happened seamlessly within the game itself by writing BASIC-like mini routines in the B.O.B system screen. However, all these perks came at a price and the game at times seemed overly complex or obscure, leaving players clueless on what to do besides forcing very frequent loading times and disk swaps.

4.10 Shoot 'em Ups

Often considered as a staple of early gaming since the appearance of Taito's Space Invaders in 1978, "shoot 'em ups", i.e. games where players usually control a spaceship or character that needs to swiftly move around while shooting and avoiding anything that moves, were plenty on the C64 as well. Anyway, even if arcade games like Galaga, Tempest, Zaxxon and others ruled the scene, C64 developers were not satisfied by simply cloning existing concepts but, instead, often used those models to craft completely original and very imaginative titles.

This was apparent from the very beginning: already in 1983 several quality productions were released starting with "Blue Max" (Fig. 4.75), a game that successfully reworked the theme of Atari's "River Raid" with Sega's "Zaxxon" isometric perspective to provide a unique shooting and bombing action set during the first World War.

That year also saw the C64 debut of one of the most iconic independent developers of all time, Jeff Minter[21] and his studio Llamasoft. Gridrunner (Fig. 4.76) and Matrix (Llamasoft, 1983) provided a compelling action by merging original ideas with elements of arcade classics like "Centipede" (Atari, 1981), "Tempest"

---

[21] Jeff Minter (born 1962) is one of the very few pioneers who are still active today, developing games for a variety of platforms.

Fig. 4.75 Blue Max (© Synapse Software 1983). Fly well and you will be ranked accordingly (from "Kamikaze Trainee" all the way up to the coveted "Blue Max" decoration) in this engaging bi-plane strafing and shootout mission

Fig. 4.76 Gridrunner (© Llamasoft/HesWare 1983). A perfect synthesis of core elements that made a few arcade games great, plus some original touches, made Gridrunner a truly addictive shooter and a small masterpiece in its own right

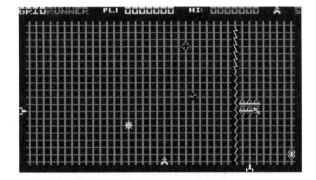

(Atari, 1981) and "Solar Fox" (Bally/Midway, 1981), and then the truly eccentric "Attack of the Mutant Camels" (Fig. 4.77), reminiscent of Parker's "Star Wars: the Empire Strikes Back", introduced one of the signature elements that became common across most of his future productions: big hairy animals such as sheeps, llamas or, like in this case, camels.

Imaginative games from Minter were a constant throughout the eighties, with "Iridis Alpha" (Fig. 4.78) probably being one of his most well known and appreciated shooters.

Being such a popular genre, it is not surprising that shooters weren't the exclusive domain of small studios founded by whizz kids like Minter but grabbed also the attention of major software houses like Activision that went on to develop very interesting products as well. Among these, "Pastfinder" (Fig. 4.79) stood out for its original gameplay that set it apart from anything released till then. Set in the far future in a radioactive world, the player took control of a four legged vehicle, the "Leeper", in a quest for retrieving long lost artefacts from an earlier civilization. The original element here was the possibility of upgrading the Leeper with different hardware, including shields, radiation absorbers or scramblers to affect enemy movements, adding a layer of tactical play that was uncommon for this type of games.

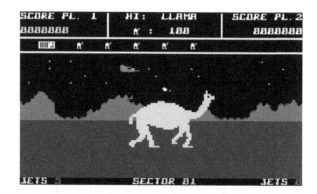

Fig. 4.77 Attack of the Mutant Camels (© Llamasoft 1983). Here the player controlled a small jet plane trying to stop a wave of giant yellow camels from reaching his home base, making this a shooter with an original tower defense flavour. Fighting each camel was a tough battle as several hits were needed to destroy her. A hyper space level in between waves was also included to add variety to the gameplay

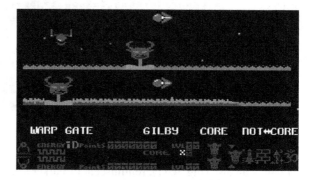

Fig. 4.78 Iridis Alpha (© Hewson 1986). Colorful graphics and surreal sound effects worked very well in this imaginative split screen shooter where players had to juggle between two ships, one for each half screen, alternating controls while avoiding crashing either one. Scoring was proportional to the ship speed, encouraging the player to take risks and travel across the split landscape as fast as possible. Surprises weren't over though, as pausing the game would start a mini game which, when paused, would also start another one!

1985 saw the release of more groundbreaking titles that experimented by mixing shoot 'em up action with other game elements: "Cauldron" (Fig. 4.80) added platforming sections and adventure overtones in a quest for retrieving ingredients of a magical spell able to defeat the evil Pumpking, while "Paradroid" (Fig. 4.81) added strategic and tactical elements to the mix. Here we played as an "Influence Device" droid in a desperate attempt to regain control of a fleet of 8 spaceships by destroying all the robotic crew members who committed mutiny before the fleet could fall into enemy's hands. A stealth approach was often needed, with the player's own droid being rather weak but able to gain control of nearby enemies for a short time by connecting to them

Fig. 4.79 Pastfinder (©
Activision 1984). While
graphics looked basic,
challenging level design
across a huge map made the
game interesting and never
boring

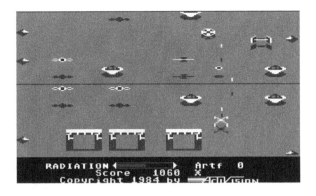

Fig. 4.80 Cauldron (©
Palace Software 1985).
Cartoonish graphics made
Cauldron come to life:
playing as a green haired
witch, the game featured
horizontal scrolling shooting
sequences on a broomstick
followed by challenging
platforming levels to
ultimately retrieve the
necessary ingredients needed
to craft a powerful spell

Fig. 4.81 Paradroid (©
Hewson 1985). Developed
by Andrew Braybrook,
Paradroid is often regarded
as one of the very finest C64
games of all time. Intuitive
controls coupled with huge,
multi-level spaceships and
24 different types of droids
guaranteed a satisfying and
compelling challenge

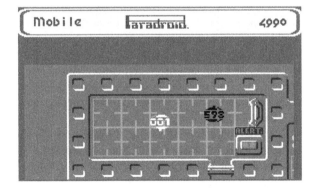

and winning a cleverly designed minigame. Once under control, the host could
be used as the player's own robot for destroying other droids and progress in the
mission.

After Paradroid, Hewson kept publishing high quality products like Uridium
(1986) but it was another software house named Sensible Software that quickly

Fig. 4.82 Wizball (© Sensible Software/Ocean 1987). Controlling a bouncing ball seemed odd at first, but the movement could be mastered with a little exercise. In any case, for particularly clumsy players, an "anti-gravity" power-up was available to let the spheres float in the air, making them much easier to control. In a single player game, players had to switch controls between the two but a two player co-op mode was also available with one player taking Wiz and the other Nifta: trying to coordinate movements as a team was a good challenge and great fun in its own right

raised to fame and stole the scene for the next few years. Masterpieces such as "Parallax" (published by Ocean in 1986), which mixed space action both on board of a small spacecraft and on foot of an artificial world, and, especially, "Wizball" (Fig. 4.82), quickly made the company become a favorite of C64 players all around the world. Like Parallax, Wizball was developed by Chris Yates and Jon Hare and featured a terrific soundtrack by Martin Galway. Anyway, it was even more original in concept than the earlier work: playing as Wiz the wizard and his cat, Nifta, the player had the mission of restoring color and happiness in the world by collecting colored droplets. These were spread across a vast world made of interconnected levels where players could move around by using two bouncing balls, one for the Wiz and one for Nifta, who had the task of actually collecting the various colors. Naturally, several enemies were moving around as well trying to stop the small team but different power-ups and extra weapons made Wiz and Nifta powerful enough to face any challenge as the adventure progressed.

The shooters to appear during the late C64 life showed once again what could be achieved on a humble 8 bit home computer, even though it could be argued they somewhat lacked the originality and creative energy of the earlier titles. Merging shooting and platforming action was getting increasingly popular in the latest arcade and console games and this template was then adopted also on the C64 without making much effort for pushing the genre further. A perfect example was Turrican (Fig. 4.83), which was an extremely polished game showcasing very high production values but lacked original gameplay ideas and instead borrowed heavily from existing games like Metroid (Nintendo, 1986) and Psycho-Nics Oscar (Data East, 1987).

Fig. 4.83 Turrican (© Rainbow Arts 1990). Turrican showed the C64 was able to deliver console quality games also in the shooter genre and it was later ported to the Mega Drive, PC-Engine and Game Boy systems. A sequel, Turrican 2, was released in 1991, maintaining, if not surpassing, the same very high standards

4.11 Sports

In his seminal work "The Art of Computer Game Design",[22] the first book ever published on the theory of video and computer games in 1984, famous game designer Chris Crawford expressed his scepticism about the feasibility and attractiveness of sports as a topic for computer games due to all the many possible factors and variables involved in the simulation, forcing developers to implement only an overly simplified and uninteresting version of the original subject. Despite his understandable pessimism, the Commodore 64 would soon prove that home computers were actually able to distil the fundamental features of competitive sports and deliver an engaging rendition of the action happening on any sporting arena.

Already in 1983, Commodore itself released what was the most realistic football/soccer game up to that day: "International Soccer" (Fig. 4.84). While many features of the actual sport like fouls, penalties, tackles etc. were still missing, the colorful sprites together with a first realistic attempt at modelling ball physics, made "International Soccer" stand out and remain the reference game till Sensible Software developed "MicroProse Soccer" in 1988 (Fig. 4.85).

Still in 1983 Gamestar released "Star League Baseball", a game that managed to recreate the popular american sport in a more realistic way than any other game before it. A side perspective allowed not only for a comprehensive view of the field but also for a more precise control of the action, like handling the height of the ball when pitching. Baseball computer games saw the next big step forward in 1985 when Accolade released "Hardball!" (Fig. 4.86), a game where not only different camera angles could be used to follow the action like a live broadcast but also plenty more options were available to deliver a true to life experience, including the choice between different players, each with their own characteristics and stats.

Basketball wasn't left behind either: following his success with "International Soccer", Andrew Spencer used the same approach for another well received game, "International Basketball" (Commodore, 1984). However, the real innovation came

---

[22] McGraw Hill-Osborne Media, 1984.

Fig. 4.84 International
Soccer (© Commodore
1983). With nine difficulty
levels, a cheering crowd and
a trophy to win, Andrew
Spencer's game was a huge
leap forward compared to
anything else seen before

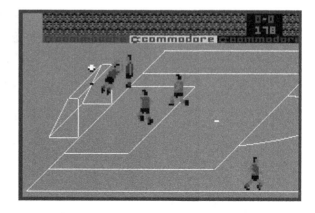

Fig. 4.85 MicroProse
Soccer (© Sensible
Software/MicroProse 1988).
A top down perspective, goal
replays, "aftertouch" effects,
different game modes,
teams and even variable
weather added a new layer of
engagement and playability
in this fast rendition of the
most popular sport in the
world

Fig. 4.86 Hardball! (©
Accolade 1985). Different
pitching options were
available while the action
unfolded like in a live TV
match. Players could also
handle some strategic and
management decisions during
the game, starting with the
selection of the team line-up
to suit a specific approach to
the game

with EA's "One on One: Julius Erving and Larry Bird" (Fig. 4.87). For the first
time, two real stars were signed to become part of the game, which was also very
well implemented with free throws, 3 pointers and slam dunks that could even break
the board in pieces! Another very interesting game was released by Activision in

Fig. 4.87 One On One
(© Electronic Arts 1984).
Playing as Dr. J or Celtics'
legend Larry Bird definitely
added a lot of appeal and
charm, starting a trend that
soon became a trademark
of games based on popular
sports

1986, "GBA Basketball". This was a two on two game where the player directly controlled only one man while selecting his team mate among several characters inspired by real athletes. Specific strategies could be called during the action making for a compelling simulation where the only major drawback was the duration of each game: unfortunately, this was set to match real events and couldn't be changed, making the playing of a whole season a marathon affaire.

Besides these classic examples, most popular sports had different games dedicated to them. Tennis fans would play "Match Point" (Psion, 1984) or pioneering 3D experiments like "International 3D Tennis" (Palace Software, 1990. See Sect. 4.14). "Face Off!" (Activision, 1987) was a great ice hockey rendition and bowling players had "10th Frame" (Access Software, 1986) to entertain them. Access Software was also responsible for another fantastic game that made any golf enthusiast go crazy: "Leaderboard Golf" (1986).

Anyway, a category that definitely had several very successful releases was fighting sports like boxing and martial arts. "Barry McGuigan's World Championship Boxing" (Sportsware Production, 1985) provided fighter customization, nice animations and intuitive controls. The game offered plenty of challenge by having players start from the very beginning of their boxing career to ultimately progress towards a world championship title: training across different disciplines was needed before being ready for a fight and, once in the ring, a wise use of available energy and stamina mattered more than simple button meshing.

Sport Karate games, with an emphasis on scoring points instead of draining the opponent's energy till a KO, had a marvellous example in "Way of the Exploding Fist" (Fig. 4.88) where beautiful oriental backdrops and music set the stage for a realistic fight with smooth animations and good variety of techniques.

"International Karate" (System 3, 1986) soon followed with a very similar gameplay that was refined even further with "International Karate +" (Fig. 4.89) in 1987, offering a frantic fighting action between 3 karatekas that provided lots of replayability also thanks to humorous touches and small easter eggs that were all to discover and enjoy (for example, a fighter could lose his pants during a fight and then stare helplessly at the player on the other side of the screen!).

Fig. 4.88 Way of the Exploding Fist (© Melbourne House 1985). Despite using only a joystick with one button, the game offered a satisfying variety of blocks, kicks and punches, all delivered with smooth animations

Fig. 4.89 International Karate+ (© System 3 1987). If a fight between two karatekas wasn't challenging enough, starting a brawl with three athletes surely was! Contrary to other games, IK+ showcased only one background but it was beautifully animated and had lots of small secrets hidden in it

Fig. 4.90 Summer Games (© Epyx 1984). The opening ceremony. Once over, up to eight players could choose their nationalities and compete across the following events: Pole Vault, Diving, 4 × 400-m Relay, 100-m Dash, Gymnastics, Freestyle Relay, 100-m Freestyle and Skeet Shooting

Multi sport events inspired by the Olympics also got very popular and several games like Activision's "Decathlon", Ocean's "Daley Thompson's Decathlon" and Epyx's "Summer Games" (Fig. 4.90) were released for the C64 in 1984. The latter was particularly good in capturing the Olympic spirit, including an opening ceremony and national anthems. The game was so successful that Epyx released

Fig. 4.91 California Games (© Epyx 1987). Summer in California is surely a great time for having fun. Besides Surfing, also Frisbee, Half-Pipe Skateboard, BMX Bike Racing, Foot Bag and Roller Skating were available to try alone or with seven other friends

a sequel with different events, "Summer Games II", besides a winter themed edition, "Winter Games", in 1985. These were then followed by two collections of less orthodox, but by no means less fun, games: "World Games" in 1986 and "California Games" (Fig. 4.91) in 1987.

4.12 Strategy

Despite being often considered as a niche genre, strategy games played an important role on the C64, especially in its early and late days when some really noteworthy productions were released. As early as 1983 Electronic Arts established its pioneering role as an innovator with its first slate of releases, which included master pieces such as "M.U.L.E." (Fig. 4.92) and "Archon" (Fig. 4.93). These were not straightforward strategy games but managed to build an engaging strategic experience by merging traditional elements with trading and economic aspects in the former and arcade, tactical action in the latter.

In 1986 Accolade proposed a very original space simulation with a strategic flavour that was much different from most other space games of the time: "PSI 5 Trading Company" (Fig. 4.94). Here players took the role of a starship commander with the task of delivering a cargo to a different planet. Crew selection was the first step of the adventure, with different roles to fill and candidates to choose, each with unique personality traits. Needless to say, many things could possibly go wrong during the cruise and maintaining the situation under control would definitely test anyone's strategic as well as people management skills.

Anyway, another landmark in computer strategy games was set 1 year later by Cinemaware with a game of territorial expansion that perfectly merged a simple, yet not simplistic, ruleset with great action sequences and a captivating historical setting: "Defender of the Crown" (Fig. 4.95). Set in twelfth century England, the game depicted the struggle for power between Saxons and Normans feudal lords, with the player taking the role of one of the former. The main game theatre was a

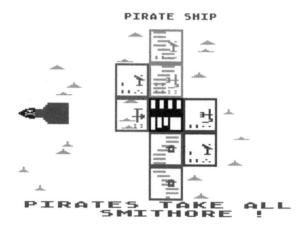

Fig. 4.92 M.U.L.E. (© Electronic Arts 1983). Robotic "mules", with the acronym actually standing for "Multiple User Labour Element", were at the centre of this colonization game where up to four players had to strategically occupy different areas of planet Irata (a joke on Atari) and mine various resources. Knowing when to compete and when to cooperate was of fundamental importance to succeed here and the game, developed, among others, by Dan Bunten, is often remembered for offering an original auction system to model and regulate its economic market

Fig. 4.93 Archon (© EA 1983). At first sight similar to chess, Archon merged pure strategy with arcade action to decide the outcome of a battle between two pieces. Each army, a Light and a Dark side, had different units with different characteristics that became apparent during the actual combats: some were strong but slow while others could fly and move very quickly. The board could also change according to a day and night cycle, giving an energy bonus to the respective side. A sequel, "Archon 2: Adept", was released in 1984

map of England subdivided in different territories but different events would play out via spectacular action sequences like jousting or storming an enemy's castle, turning the game into a cinematic experience that was Cinemaware's own signature.

Jumping to another, much more recent, historical setting, in 1988 Commercial Data Systems offered "Tank Attack", a very interesting adaptation of a classic war game from 1977 with the same name. However, this wasn't a standalone computer

Fig. 4.94 PSI 5 Trading Company (© Accolade 1986). Detailed information on the status of the starship and presence of any eventual nearby spacecrafts were always available and were important elements in deciding the proper course of action to avoid ending stranded somewhere in space

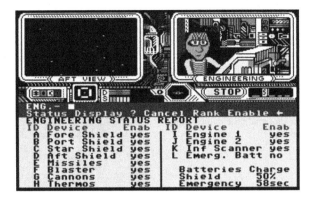

Fig. 4.95 Defender of the Crown (© Cinemaware 1987). 12th century England. Notice the presence of Sherwood forest: in case of emergency, the player would even adventure over there to seek the help of an old friend, the legendary Robin Hood!

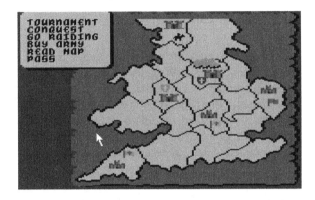

game but a sort of hybrid: the game box, besides the manual and tape (or floppy disk), also included a 40 cm × 40 cm board and 48 plastic pieces representing tanks and other armoured units. The game needed two to four players, with each in control of a small fantasy nation with names like Sarapan and Calderon, and its ruleset was relatively simple, making for a good introduction to tabletop games. In this context, the computer tried to bring additional excitement, for example by providing news-feeds updating the players on the war status and, most importantly, acted as a sort of moderator handling different events. These included battle outcomes which were determined by taking into consideration not only the forces involved and their relative position, but also to a few external factors like terrain and weather conditions.

Moving from the conquest of nearby countries to the colonization of a whole solar system, Melbourne House in 1991 released one of the last great strategy games on the C64: "Supremacy" (Fig. 4.96), also known as "Overlord" in the US. Four levels of difficulty were available, each affecting the scale the solar system and represented by a specific opponent who belonged to different alien races. These ranged from strong but stupid to almost perfect in every aspect of strategy and warfare. Gathering resources, building units and finding the right taxation point to earn money why also keeping the population healthy and happy, were all of paramount importance to succeed. At the same time, the different approaches needed to defeat the aliens in the fight over the planets guaranteed a good variety to the resulting gameplay.

Fig. 4.96 Supremacy (©
Melbourne House 1991).
Excellent graphics and
music (by Jeroen Tel) made
Supremacy a pleasure to play,
even though it included an
almost 100 page manual that
definitely had to be studied
to appreciate the game to the
fullest

4.13 Virtual Life

Even though virtual life computer games reached the mainstream public only
recently thanks to Will Wright's PC game "The Sims" (Electronic Arts, 2000) and
its groundbreaking success, their origins can be traced back almost two decades to
a computer game that, regrettably, has been forgotten by many: "Deus Ex
Machina",[23] published in 1984 by Automata UK on the ZX Spectrum and then
ported to the C64 by Electric Dreams. Designed by Mel Croucher, this was something completely new, as described in its manual:

> A completely new form of computer entertainment. You Play the leading character in this
> fully animated televised fantasy, controlled by your home computer and synchronized to
> its own stereo soundtrack. It is a union of computer game, film, book and L.P. record. It is
> the first of a new era of experiences and it is unique. Enjoy it.

Indeed, this was not just a "game" but a complex multimedia performance, a the-
atrical piece even, where the onscreen action had to be accompanied by a music
tape including narrative. The gameplay idea behind "Deus Ex Machina" was to
follow the development of a new, computer generated human life with the player
tasked to follow his whole development, from conception through evolution and,
ultimately, death.

The progression, loosely based on the "Seven Ages of Man" from Shakespeare's
"As You Like It", happened via different mini games where the player had to perform some simple task in a straightforward arcade style action (Fig. 4.97).

While way ahead of its time, "Deus Ex Machina" was also a very linear experi-
ence where players could only follow the predefined path laid down by its author.
To actually be in charge of their own "virtual" destiny, players had to wait for
"Alter Ego", released in 1986 by Activision. Designed by a psychologist, Dr. Peter
Favaro, the game offered both a male and a female version where players could,
once again, go through all the "seven ages of man" by advancing on a branching

---

[23] A successful crowdfunding campaign on Kickstarter to bring back the game to modern platforms was launched by Mel Croucher in 2014.

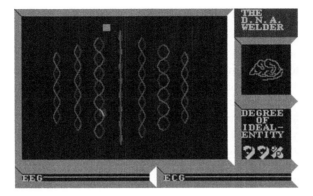

Fig. 4.97 Deus Ex Machina (© Automata/Electric Dreams 1984). The "DNA Welder" at the beginning of the game, where the player has to keep the DNA strings spinning so that the machine can analyz them to create a new, unique individual

Fig. 4.98 Little Computer People (© Activision 1985). We could remind our little friend to keep fit, practice the piano or play computer games, among many other things to take care of his well being. If happy, or if something needed to be improved, he would sit down and write us a letter, giving valuable feedback besides sharing his feelings with us

tree where each leaf represented a different event. These included health, work, relationship and other topics, offering players different actions, each potentially yielding long lasting consequences and providing great replay value.[24]

In any case, the closest we could get to the modern concept of "virtual life" as established by EA's "The Sims" was in "Little Computer People" (Fig. 4.98), another groundbreaking game published by Activision. Starting from the assumption that some little "creature" was actually living inside our computer, the game depicted the ongoing life of one of these inhabitants and his dog. As their host, we could communicate with them by typing in simple commands and help them in living a good life, establishing a friendly relationship that was actually fun and emotionally fulfilling.

---

[24] The game has been ported to several platforms in recent years and can also be played online at http://www.playalterego.com/.

4.14 3D: Vectors and Polygons

The Commodore 64 wasn't build to provide fast 3D based graphics, nonetheless it is undeniable that the third dimension always attracted developers and made them constantly try to push the hardware beyond its limits. Already in 1983 Penguin Software managed to propose a pretty impressive game, "Stellar 7" (Fig. 4.99). Clearly inspired by the coin-op Battlezone (Atari, 1980), the game opened spectacularly with the player's own tank moving out of a hangar and revealing a barren alien world, starting a fight across different planets on seven solar systems.

No early computer game, though, managed to impress and excite people by implementing 3D graphics into an open universe full of possibilities like "Elite" (Fig. 4.100), originally developed by David Braben and Ian Bell for the BBC Micro in 1984 and then ported to the C64 in 1985 by Firebird.

Fig. 4.99 Stellar 7 (© Penguin Software 1983). Tank, spaceship and turret designs were extremely basic but the wireframe graphics outlining very simple 3D models worked well enough to bring its virtual world to life

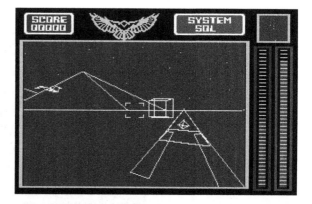

Fig. 4.100 Elite (© Acornsoft/Firebird 1984/1985). Players were graded throughout the game, starting as "Harmless" and then progressing on the ladder of fame and fortune up to the coveted ranking of "Elite". Not many people achieved it

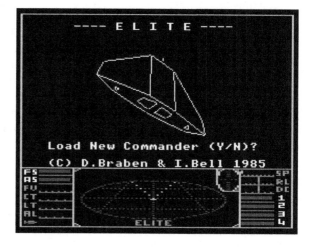

Fig. 4.101 Mercenary (©
Novagen 1985). The world
of Targ wasn't densely
populated on the surface.
On the other hand, many
things were happening
underground where one of
the two warring factions was
hiding. The action could take
place anywhere: on, above
and under the surface of the
planet

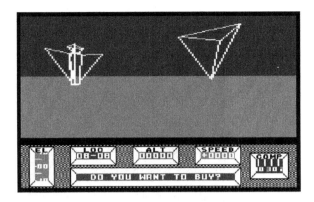

Set in a galaxy with more than 2,000 unique planets, each with its own characteristics, economy and alien culture, players were free to roam space being whatever
they liked: a peaceful trader, a savage pirate or anything in between. Ships could be
upgraded and customized with additional weapons or cargo storage as needed and,
while standard missions were also available, the game gave players complete freedom to plan their own adventures and routes, marking the birth of the open world
genre that was destined to become so successful decades later.

Another game that gave players plenty of freedom was "Mercenary" (Fig. 4.101),
released in 1985 by Novagen. Stranded in the world of Targ, where two factions
were at war against each other, the player could side with either party in a quest to
gain enough money to buy (or steal …) various equipment to prepare a new spaceship and travel back home.

Anyway, combat simulations didn't have to be always set in space or in some
alien world and MicroProse released "Gunship" to critical acclaim in 1986. This
was a very detailed military simulation where players controlled an Apache combat helicopter across different missions and scenarios spanning South East Asia,
Europe and more. Controls were complex, with many different weapons to choose
from according to specific targets, but the 3D wireframe graphics worked really
well, effectively using colors to identify different terrains as well as time of day.

3D Games didn't have to be all about warfare either, as demonstrated by Firebird
with its 1986 classic "The Sentinel" (Fig. 4.102), a strategy game that defied classifi-
cation,[25] or by the pioneering tennis simulation from Palace Software "International
3D Tennis" (Fig. 4.103), one of the first sport games that traded nicely animated
sprites for a more technologically advanced look and feel.

"The Sentinel" was also notable for starting to fill the wireframe graphics with
color, turning otherwise transparent objects into proper 3D solid models. The
game didn't really need moving graphics, though, so when "Driller" (Fig. 4.104)

[25] Indeed, Zzap!64, while awarding it a "gold medal" with comments like "the best game ever
written for a computer", didn't rate it with a numerical score (issue 20, December 1986, p. 22).

Fig. 4.102 The Sentinel (© Firebird 1986). Designed by Geoff Crammond, it offered ten thousands engaging levels, each being a unique puzzle where the player needed to defeat an existing sentinel by moving undetected across the 3D landscape and, ultimately, absorb its energy. The energy system was a truly unique feature of the game, with the player and the sentinel absorbing and releasing it from the environment to alter the landscape and fight against each other

Fig. 4.103 International 3D Tennis (© Sensible Software/Palace Software 1990). Offering a single match, a tournament or a full season with different surfaces to play on and different camera angles, the game had plenty of options to keep gamers interested. And the 3D look definitely added something new to the overall experience

was released 1 year later in 1987, many jaws dropped in awe. The game was similar in scope to "Mercenary" but pushed the existing technology one step further with its 3D filled polygons and, while the screen refresh rate was very slow by modern standards, it was actually fast enough to make the game playable and, most importantly, enjoyable.

Ideally closing the circle, the last great 3D game on the C64, "Battle Command" (Fig. 4.105) went back to the "Battlezone" template like "Stellar 7" did in 1983, but re-imagined the experience showcasing all the technical and design knowledge acquired during those years. Not only polygons were now filled but actually moved fast, the battle field was varied and the game offered a good variety of different missions. Several small touches also added realism and immersion, like the weapon selection screen featuring 3D model of the actual weapon spinning around for a better view, or the possibility of using binoculars and a night vision mode

Fig. 4.104 Driller (© Incentive Software 1987). Set on the distant moon of Mitral, a build-up of gas under the surface is threatening the very existence of the planet. The player is then tasked to explore the deserted, but still dangerous, moon in an excavation probe to find gas pockets and place drilling rigs to drain them before they could explode. Exploration, environmental puzzles plus some combat action and a haunting 15 min music track by Matt Gray, sealed the deal and established the game as one of the most impressive production of the late 1980s

Fig. 4.105 Battle Command
(© Realtime Games/Ocean
1991). Released first on
the Amiga, many people
doubted the C64 could
handle such a complex and
computationally intensive
game but they were soon
proved wrong!

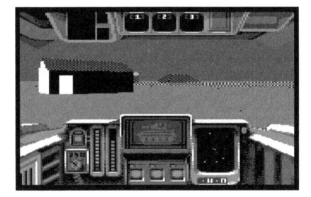

while in the heat of the action. Once again, the old C64 ended its official life on the shelves as a gaming system with something so impressive that most people would have thought to be impossible.

Chapter 5
Rise of the Game Engines

Almost all games developed today are not built from scratch but by using powerful middleware tools, commonly named "game engines". These provide programmers, artists and designers with a set of commonly used features and systems already implemented and ready to be reused to craft an original concept, considerably speeding up the development process. On the other hand, back in the 1980s, most games were developed in assembly language starting from zero and, while the various computer architectures were much easier to understand and master than modern counterparts, this barrier still made game development hard, especially for non technical people. Nonetheless, being creative and able to express ideas by writing their own programs and games, was always one of the main reasons or, sometimes, main excuses, used by kids to ask parents for investing in one of those new 8-bit technological marvels. This hunger for creativity didn't pass unnoticed and it wasn't too long before some of the leading developers in the newborn computer game industry decided to release the first tools offering users a shortcut for simplifying game development.

1983 was the year when the first generation of what we could call today "level editors" and "game engines" were released for the general public. Douglas Smith's "Lode Runner" (Broderbund) and Bill Budge's "Pinball Construction Set" (Electronic Arts) were the first titles to allow players, on the C64 as well as on other platforms, to make their own original levels or, in case of the latter, a complete pinball table. Kids of any age were now able to extend the fun and replay value of those games in an original way, implicitly learning the skills of level design. Anyway, making a fully fledged game engine able to simplify development for almost any kind of game, like those we have available today, was obviously well beyond the reach of the technology of the time and the first commercial products had to limit themselves to very specific genres only.

Still in 1983, John Hollis wrote the "Games Designer" (published by Software Studios) on the ZX Spectrum for making simple arcade games in the style of Galaxian or Asteroids while Gilsoft released "The Quill", first on the ZX Spectrum and then on the C64 soon afterwards, for creating original text adventures. Programmed by Graeme Yeandle, "The Quill" actually allowed its users to release their games commercially and was fundamentally responsible for starting

© Springer Science+Business Media Singapore 2015
R. Dillon, *Ready*, DOI 10.1007/978-981-287-341-5_5

the irresistible rise of game middleware that kept lowering the barrier of entry to the game development world till this very day.

5.1 The Quill—Adventure Writer (Gilsoft 1983)

Before the "The Quill", game designers like Scott Adams[1] already had developed their own reusable game engines to handle databases and the parsing functionalities necessary for writing adventure games. While they did not share their tools as stand-alone commercial products, their efforts did indeed spark the curiosity of the hacking and programming communities: reverse-engineering attempts of existing games and related articles to understand how such systems would work, started appearing on computer magazines and gradually became more and more insightful. It was in this creative and exciting hacking environment that the Quill was born.

The program, which occupied a mere 12 KB of memory, booted by presenting a menu outlining all the options needed for defining the game, starting from a "Vocabulary" where to input all the words required. Once this was done, the budding designer would proceed to the "Location Text" and "Object Text" menus, which allowed for the definition of the actual descriptions that were to be displayed during the different phases of the game. Movement between locations and object placement was handled by other menus ("Movement Table" and "Object Start Location" respectively) while events, like the control and manipulation of objects in the inventory to solve puzzles, were handled in the "Event Table".

The different pieces of the adventure had to be patiently put together one by one via the various tables so, before embarking into such a work, the game had to be completely designed on paper with all words, locations, text and puzzles carefully planned in advance.

The Quill was not only a groundbreaking product and idea in itself but was way ahead of its time also in its licensing terms: anyone was free to release their games commercially, with only a mention in the credits being appreciated. Such permissive terms were unheard of at the time and turned it into an overnight success, making it the design tool of choice for any aspiring adventure game programmer.

In the end, several hundred commercial adventure games were released and localized in different languages across the various platforms the Quill was ported to.

5.2 The Graphic Adventure Creator (Incentive Software 1986)

Despite their early popularity, text only adventures soon started to show signs of fatigue and images accompanying verbose descriptions became increasingly common. To avoid being left behind, Gilsoft released a Quill add-on, "The Illustrator"

---

[1] Scott Adams (born 1952) developed some of the most successful early text adventures on home computers like the TRS-80 and the VIC-20, published by his own company "Adventure International".

to add drawings to its games. The same approach, merging text and graphics but in just one complete package, was also implemented by another game engine that tried to improve over "The Quill" by adding also a few other important features: "The Graphic Adventure Creator", designed by Sean Ellis and released by Incentive Software in 1986.

Following the successful approach of the Quill, the program started with a main menu (Fig. 5.1) where several options were available. Users could define a room description, specify the connections between different locations, define different objects, specify the messages a player could get when entering a room or performing an action, adding conditions (a switch could be on or off, for example) and more, besides accessing also a graphic program (Fig. 5.2) where to draw lines, dots and various shapes using different colors.

Fig. 5.1 The main menu of the "Graphic Adventure Creator". The whole adventure had to fit in 23,113 bytes only

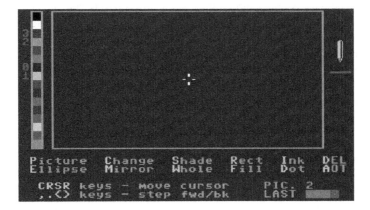

Fig. 5.2 The built in drawing program. Despite the very simple tools offered, it was possible to craft some relatively impressive images, on par with most commercial adventure games of the time

Fig. 5.3 Imagination (© Firebird 1987) was one of the commercial titles made with the Graphic Adventure Creator by a major software house

The ability to use conditions was an important step forward in game design terms compared to the Quill since it allowed for more complex phrases than simple verb plus noun sentences: for example, commands like "GET THE LAMP THEN LIGHT IT" were also possible since the lamp could now be in either an on/off state that could affect the development of the story.

The Graphic Adventure Creator was notable also for its effort in teaching users by providing not only an example game to showcase the software capabilities but also a "Quickstart" data set including a common collection of verbs and messages to speed up early development and make the learning curve smoother.

Last, if all these features were not enough, Incentive Software also decided to follow the Quill footsteps regarding user generated content: competitions were arranged where users could submit their games for a publishing agreement (which included a 10 % sales royalty) but anyone, indie developers or bigger companies alike, could release their games independently. The only requirement was to include the "Graphic Adventure Creator" in the credits.

Once again, more than one hundred games were commercially published across the various computers the engine was released for, including games by well known labels like Mastertronic, Firebird (Fig. 5.3) and CodeMasters.

5.3 Adventure Construction Set (Electronic Arts 1984)

After the success of text adventure engines like the Quill, it didn't take long to have software targeting more action oriented games. Developed by Stuart Smith, Electronic Arts soon released the "Adventure Construction Set" to let people design their own Ultima-like RPG games (Fig. 5.4).

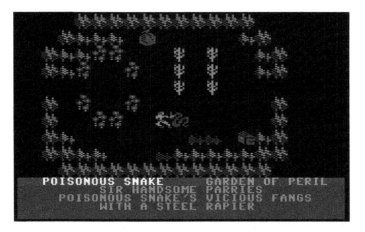

Fig. 5.4 One of the demo adventure games included in Stuart Smith's "Adventure Construction Set"

The program granted budding designers a lot of freedom while also trying to simplify the creative process as much as possible: three different "moods" were available to start with (fantasy, sci-fi and spy/mystery) and then the game could be developed by moving across an articulated system of menu screens (Fig. 5.5). Here, besides basic information like Title and Author, it was possible to structure the gaming world across three different resolution levels. The main, overall map was simply named "World" and could be subdivided into a set of "Regions" (up to 15 different ones) which could then be broken further down into "Rooms" (up to 16 per region). Following this approach it was actually possible to design worlds big enough to be varied and challenging. Characters were defined in terms of several stats like Speed, Power, Wisdom, Strength, Dexterity and more while their graphical appearance was borrowed from other EA games like Archon, MULE, Realm of Impossibility etc. but images were static and lacked animations. In any case, it was possible to assemble a party of up to four characters, allowing for interesting multiplayer adventures with the game progressing in a turn based fashion.

The Adventure Construction Set was a very ambitious project, maybe even too ambitious since some of the procedures were definitely cumbersome: for example, newly made adventures had to be stored on an dedicated floppy that had to be pre-pared beforehand via many disk swaps.[2] Besides, games couldn't be played as standalone products but always required a copy of the original software to be loaded, effectively making distribution impossible. To make up for this, at least partially, EA did organize a game creation competition with prizes up to $1,000 for the best user generated adventures.[3]

---

[2] This was something that really angered Steve Cooke, the adventure games reviewer on Zzap!64 under the screen name of "the White Wizard", who gave the program a terrible 24 % score (issue 7 1985, p. 82)!

[3] About 50 games were submitted.

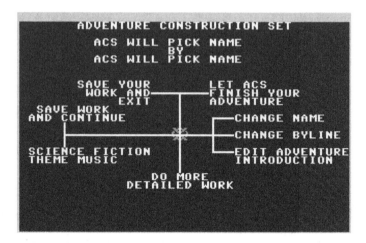

Fig. 5.5 The first menu screen starting the adventure creation process. Notice the "Let ACS Finish Your Adventure" option that would have the program automatically complete, or even build from scratch, a game with the user having only to specify parameters like the difficulty level (from 1 to 10), the final objective (like finding a specific treasure) and any eventual region attributes. Having the computer build a game from scratch would have taken about 40 min, with the C64 constantly pushing messages on the screen like "Adding Forests And Seas" or "Exploring Map" to keep the player updated during the long generation process

In addition to EA's effort, another engine is worth remembering here: the "Adventure Creator" released by Spinnaker in 1987. This was the only engine that had a cartridge release besides the common tape and floppy versions and was dedicated to simple action adventures and dungeon crawlers. Action was fast and characters were moving in real time with nice animations but, unfortunately, while it avoided the disk swap issues that marred the original "Adventure Construction Set", it also lacked its ambition and variety and remained a niche product unable to impose itself in the market.

5.4 Garry Kitchen's Game Maker (Activision 1985)

Since arcade style action games were by far the most common genre in the 1980s, it's no wonder different game engines were released targeting the development of this type of games as well.

Taking over the legacy of the original "Games Designer" from Software Studios, Mirrorsoft released the "Games Creator" in 1984, developed by David and Richard Darling, two young brothers who were soon going to start their own company, CodeMasters, with fantastic success. A few years later, Broderbund also published the "Arcade Game Construction Kit", a similar tool developed by Mike Livesay. However, the software that became most popular was "Garry Kitchen's Game Maker" released by Activision in 1985.

Fig. 5.6 The "Sound Maker" module that allowed easy access to the SID chip functionalities for shaping any in-game sound

Like other toolkits, Game Maker was also organized into a set of utilities that complemented and integrated each other to produce a final game. In particular, here there was a "Scene Maker" for drawing background graphics (with one main background color and three others), a "Sprite Maker" for in-game objects, a "Sound Maker" module for shaping SFX that looked very similar to an actual synthesizer console (Fig. 5.6), and a "Music Maker" program for composing melodies by writing notes on an actual musical score (Fig. 5.7). All these were joystick controlled with every function accessible via easily understandable menus.

All the assets prepared could then be accessed and put together to build an actual game via the "Editor" window (Fig. 5.8). This was an actual programming environment where simple BASIC-like instructions could be used to code the game, issuing commands for playing songs, displaying sprites in certain positions etc.

The working environment set up in Game Maker was truly impressive and way ahead of its time, even though there still were heavy limitations, mostly due to the lack of available memory: only two stationary backgrounds could be added, meaning that no smooth scrolling across areas bigger then the screen was possible, and, most importantly, only 3553 bytes were actually available for game resources, including assets and code.

Despite these problems, this was probably the most flexible engine ever developed for the C64, actually able to produce a wide variety of different small games[4] that could also be saved and run independently from the original tool.

---

[4] Activision also released a Sports and a Science Fiction add-on featuring additional resources including graphics and sounds suitable for those types of games.

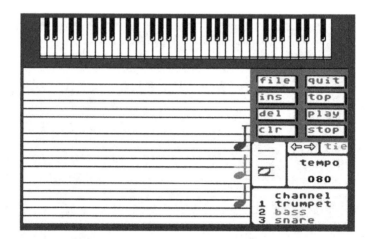

Fig. 5.7 The music composing tool

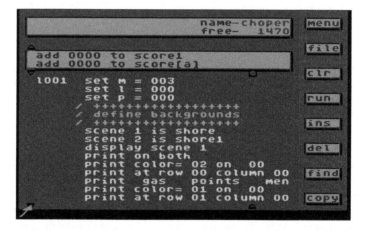

Fig. 5.8 The actual coding editor window where games were built by using a BASIC like high level scripting language

5.5 The Shoot 'em Up Construction Kit (Sensible Software 1987)

Among the different genres that had their specific game engines, we shouldn't forget SSI's "Wargame Construction Set", released in 1987. This was a very comprehensive take on wargames that allowed for both small and large scale strategic battles across a variety of settings, from ancient to modern and science fiction. Nonetheless, the engine that became by far the most popular was dedicated to another type of games: vertically scrolling shooters and it was named "The Shoot 'Em Up Construction Kit", often referred to simply as SEUCK.

Released by Sensible Software in 1987 and met with instant critical acclaim ("a milestone in computing history" wrote Julian Rignall in the December 1987 issue of Zzap!64), it soon became one of the most beloved program to design games, so much so that it is still popular nowadays with a small but enthusiastic following of developers who keep releasing new games made with it.[5]

SEUCK, like any other engines discussed earlier, was also subdivided into different small utilities with each one building a specific component of the final game. This could then be run as a standalone program for the delight of all the programmer's friends.

Due to its historical relevance and continuous popularity,[6] it seems appropriate to discuss this engine a bit more in depth and see how we can actually start developing a simple game with it via an introductory step by step tutorial. The reader who doesn't feel like tinkering with emulators or real hardware at this time may safely move to the next chapter and leave the following primer for later study.

5.6 A SEUCK Primer

After loading SEUCK we are presented with a main menu featuring different options as shown in Fig. 5.9.

Let's start by selecting "**Edit Sprites**" and then, in the following menu, choose "**Select Sprites**" (press **F1** or the joystick fire button with the menu option highlighted). Here, by moving the joystick up and down, we can see all the sprites than have been defined so far. By default, Sprite 000 looks like Fig. 5.10: a simple triangle with a shadow effect for modelling a very basic spaceship.

By pressing the joystick button (or the keyboard shortcut **F3**) we can then proceed to edit the currently visualized sprite and a cursor will appear inside the sprite grid at location 0, 0. Notice how the grid is only 12×21 since we are dealing with multicolor sprites here: the higher resolution 24×21 grid we discussed earlier in Chap. 3 applies only when drawing monochrome sprites. In any case, besides the resolution, there are some other restrictions we have to live with here: only four colors are possible and the first three are also shared between all sprites, meaning that we need to plan very carefully for the color palette we would like to use in our game. Note how the cursor can be freely moved across the grid: both the color list on the left and the global color palette at the bottom of the screen are accessible now to pick and assign the specific ones to be used.

---

[5] For example, information and new games made for the annual TND competitions organized by Richard Bayliss can be found here: http://tnd64.unikat.sk/.

[6] A modified version to design horizontally scrolling games was developed in 2008 by Jon Wells (more information can be found here: http://gamesplaygames.co.uk/seuck/) while Martin Piper recently developed a new game engine, named "SEUCK Redux", able to use SEUCK data (check http://www.seuck.retrogaming64.com/redux.html) and add many new features and performance improvements to the resulting games.

Fig. 5.9 The main menu in SEUCK. Every option is accessible either by joystick movement or via keyboard shortcuts. Remember that pressing X always goes back to the main menu while pressing Spacebar in a submenu goes back to the parent menu

Fig. 5.10 Default Sprite 000. We can define up to 127 different sprites to build up game characters and their corresponding animations

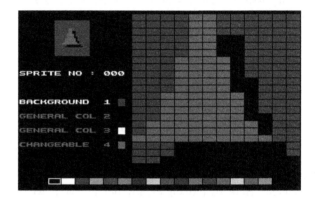

To start drawing our own sprites, press **E** to erase the default spaceship and then proceed in setting up the color palette you desire.

For this exercise I'll redraw the main character for "Shaken", a small game I developed for the TND SEUCK 2014 Competition: a fearless ninja.[7] We will be using dark grey as Background 1, black as General Color 2, Pink as General Color 3 and a Light Grey as Changeable Color 4. Once done, draw a front view the main character, like in Fig. 5.11.

We can then copy this sprite to Sprite 001 and then flip it horizontally to have a second frame we will be using for the walking down animation. To do this, press **C** and select the sprite you want to copy from via the joystick first. Choose it by pressing Fire and then select the destination sprite (001 in this case). For mirroring, press **M** and then, via the joystick, control how you want the image to be flipped. Once done everything correctly, we should have an image like the one in Fig. 5.12.

We have now to define the frames for walking up, left and right. All frames will be put together later by using the "**Edit Objects**" menu option and animations for each direction will be composed by two frames only.

---

[7] The game can be downloaded here: http://tnd64.unikat.sk/seuck/SEUCKCompo2014/Shaken.zip.

Fig. 5.11 Drawing the *front view* of the player controlled ninja in Sprite 000

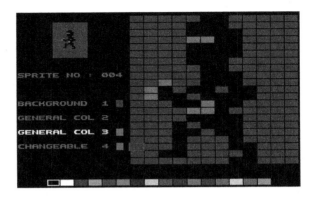

Fig. 5.12 Sprite 001 is a copied and mirrored version of Sprite 000. We are going to use this for the walking down animation

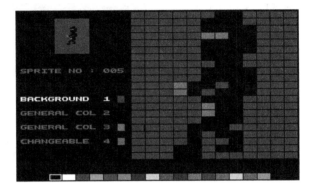

For walking upwards we can easily adapt our original sprite 000 once again: copy it in 002 and then simply turn the pink pixels black to simulate the back of the ninja hood. We can also add some grey pixels on the back (Fig. 5.13) to differentiate it further from the frontal view if we wish to do so: with some imagination, we can think of this like our hero's family crest embroidered on the ninja suit!

Now copy sprite 002 into sprite 003 and mirror it like we did earlier to have the second frame for the walking up animation.

Walking left and right requires two complete new drawings. Select Sprite 004 (if there is some sprite already stored, erase it by pressing **E**) and start drawing the first frame for walking left. Draw the second frame in Sprite 005. Possible outcomes are shown in Figs. 5.14 and 5.15 respectively.

Now copy Sprite 004 and 005 into 006 and 007 and mirror them accordingly for the walking right animation.

We can now leave the sprite drawing section and actually build the player object with all its corresponding animations. To do so, press **X** to go back to the Main Menu and then **O** to enter the Object menu.

The Shoot 'em Up Construction Kit allows for the definition of as many as 58 objects to store the actual animations of player characters, bullets, enemies and death sequences. For simplicity, they are already assigned initial values and names, so Object 00 is going to be the "**Player 1 Ship**" (i.e. ninja, in our case).

Fig. 5.13 Sprite 002 is going to be used as the first frame for walking up animation

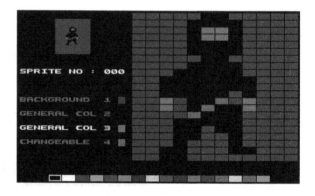

Fig. 5.14 Sprite 004, first frame of our walking left animation

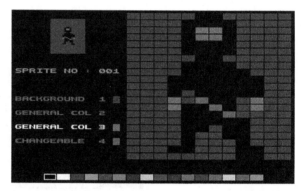

Fig. 5.15 Sprite 005, the second frame we will be using for the walking left animation

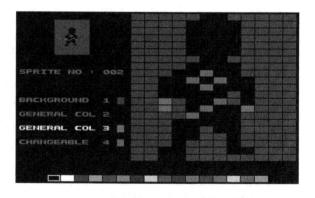

Press **F1** to move between all possible objects if you like. By default, the player object 00 is set to have a linear animation (which can consists of up to 18 frames) but that's not what we need for our Ninja, whose animation must be controlled by joystick input. To change this, press **T** and then scroll across all options: choose "Directional Hold" ("hold" means the object will stop on its last played frame when the joystick is released, otherwise the two central frames will loop).

The display will change showing a set-up for an 8-directional movement with 2 frames for each direction as discussed earlier.

Fig. 5.16 Defining all animation frames for Object 00, i.e. the Player

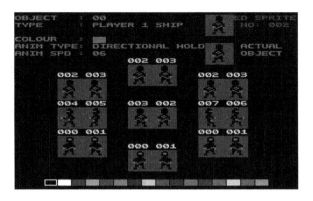

Now, to start adding the frames, we need to select the "**Edit Sprite and Place**" command by pressing **F7**. We can then choose a specific sprite via the joystick, select it by pressing Fire and then place it in the specific animation position.

Once all sprites have been placed in their respective frame slots, we should have set up an animation for the player as shown in Fig. 5.16.

We can test the animation right away by pressing **F3** and control the character via the joystick: our small ninja within the "Actual Object" square will move accordingly. We may also want to set up a proper animation speed by pressing **S**. possible values range from 1 to 16, with 1 being the fastest.

Having the main actor ready for action, we can now start drawing the background where the adventure will take place. This is defined by an overall "map" made by combining together a set of possible "blocks". Each block is, in turn, made by a 5 × 5 grid of "chars", with each of this made by 4 × 8 pixels which we can draw like we did for sprites. Up to 254 chars can be defined and, like sprites, each char can showcase four colors where the first three are common across the whole map and only the fourth (picked from the first eight colors in the palette only) is specific to a given block.

Drawing a beautiful background is, while definitely possible, one of the hardest tasks in SEUCK: it is a bit like drawing a big picture, break it into many small pieces and then rebuild it like a jigsaw puzzle.

For the sake of this tutorial, we will keep the background design to a minimum. Exit the Object section by pressing **X** and then press **B** to get into the background menu followed by **S**. SEUCK will now display all the available blocks that can be used to build up the map. Select the first one (or any other you like) and hit **F3** to start editing the chars and put together the chosen block.

The screen will change to something similar to Fig. 5.17: the previously selected, and now editable, block is displayed on the lower right corner of the screen with all available chars shown at its left. The color palette is at the bottom while the upper half of the screen shows the selected colors besides a zoomed in view of the currently highlighted char for editing.

Note that, while in the background screen, we can use the joystick to move across all areas (i.e. char editing, char selection, color palette and block editing).

Fig. 5.17 The background drawing screen. All the 254 chars displayed in the *lower half* of the screen. The char being highlighted is ready for editing in the *upper left corner*. Chars can then be selected to design the *block* shown on the *right*

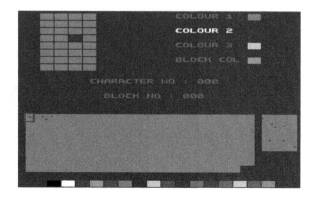

In any case, besides joystick movement, keyboard shortcuts are always available and we can quickly switch to selecting a specific char via **F1**, edit it via **F3**, change the color palettes via **F5** and **F7** and so on.

For the scope of this tutorial we want to simply simulate a plain grass surface, possibly with some brown spots to add a little bit of variety. To do so, start by selecting an appropriate color palette having dark green and and brown among its base colors. Once done, edit the first few chars by adding a brown spot here and there. If you would like to design a more complex background, with trees and rivers for example, press **S** to go back to the block selection screen and pick a different one. Once the desired block is selected, go back to the char editing screen by either by pressing **F3** or the joystick button: the newly chosen block will now be displayed on the lower right side of the screen ready to be edited by combining any available chars.

Once done, press **M** to move to the Map drawing screen. Here, by using the joystick, we can finally design the game map by placing a block at a time. In this case the map will be extremely basic, showing only one grass and dirt block repeated all over (Fig. 5.18) but, for complex maps that use different graphics, any time we need to change block we have to press **B**, going back to the block selection screen, move the highlight cursor on the desired one and then hit **M** once again to go back to the map for placing it anywhere we like.

We can now go back to the main menu and then test the game by pressing **T** and then **F1**[8]: we should have our animated ninja able to run around the screen!

Press either X or the space bar to exit the game and go back to SEUCK. It is now time to add the ninja stars so that our character can actually shoot something.

Go back to the **Edit Sprite** menu (press **S**) and find two empty slots (press **F1** and scroll through all sprites, edit the selected sprite by pressing **F3**) where to draw the frames for a very basic rotating star animation (for example, in the shape of a '+' and an 'x').

Once done, move to the **Edit Object** section and press **F1** to select object 01, named "**Player 1 Bullet**". Press **T** to select the animation type (let's choose

---

[8] Pressing F3 would start a cheat mode giving us infinite lives.

Fig. 5.18 A simple grass
and dirt background made up
by repeating the same block
previously drawn in Fig. 5.8

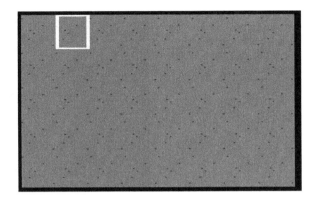

Directional) and then **F7** to select and place the previously drawn frames. In the
end, we should get something like Fig. 5.19.

If you like, go back to test the game and see that our ninja can actually shoot
now!

Now, before proceeding in adding an enemy, it would be a good time to final-
ize our ninja's abilities and actually define the level characteristics. From the
Main Menu, press **P** and then **F1** to enter the **Player 1 Limitation** screen. Here
(Fig. 5.20) we can review and modify all important data concerning our player:
number of lives, speed, how far bullets can go etc. Of particular importance is the
"**Collision with Char**" parameter: any block built with chars labelled with an ID
equal or higher than the one specified here will be impassable to the player. This
means that blocks representing obstacles, like walls or rivers for example, should
be composed by chars identified by a higher ID while roads, grass and so on
should be made with chars having a lower identification number.

The "**Edit Player Area**" and "**Edit Start Position**" allow for defining how far
the player can move around the screen and where exactly he is going to start his
journey (Fig. 5.21).

Once back to the **Main Menu** screen, pressing **L** brings us to the "**Edit Levels**" submenu with two editing options: "**Level Parameters**" (**F1**) and "**Level Map**" (**F3**).

Fig. 5.19 Animation frames
defining the "Player 1 Bullet"
object, i.e. a throwing star

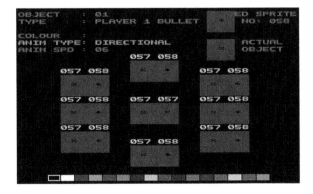

Fig. 5.20 The Player 1
limitation screen, where
many important game
parameters are summarized
together

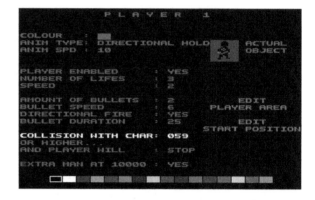

Fig. 5.21 The "Edit Start
Position" screen. Here
we can place the player
sprite anywhere within the
boundaries set in the "Edit
Player Area" screen, allowing
the player to move anywhere
within the boundaries
identified by the star-like
markers

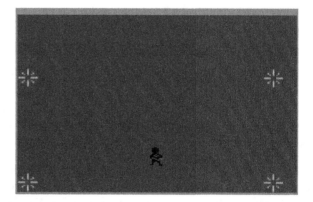

Pressing **F1** opens the Parameters screen. Here we can define how many levels are going to be included across the overall map. Each level has its own set of
parameters and can behave differently from the others: levels marked as "**Scroll**"
would have the background scroll automatically at the selected speed while levels
labelled "**Push**" would need the player to go the upper edge of his movable area to
trigger the scrolling effect. Levels could also be marked as "**Still**", i.e. without any
sort of scrolling. This could be desirable in specific situations, for example to feature a boss battle, but, in this case, a time limit must be set after which the game
progresses automatically to the next level.

Selecting a level and then pressing **F3** gives access to the overall game map
where two joystick controlled cursors could be moved to define the lower
and upper limit of the level, effectively subdividing the map into contiguous
subsections.

We are now ready to add some challenge by designing an enemy equipped to
attack the player. For simplicity, let's have an evil spear-throwing ninja, so that we
can partially reuse our existing ninja sprites.

Let's draw a few frames of the enemy walking forward and then add them
together via the **Edit Objects** menu, this time choosing a linear animation

sequence (press **T** to select it), matching the number of frames we drew previously, like in Fig. 5.22.

The spear can be designed very easily as a straight black line with a silver tip and its single frame imported in one of the available **Enemy Bullet** objects (Fig. 5.23). Now, with the **Actual Enemy** object displayed, we can put the two together by pressing **E** to get into the "Edit Enemy Bits" option (Fig. 5.24). This screen is very important as it allows us to define all important parameters for the enemy (hits needed to kill, points awarded etc.) as well as start modelling his

Fig. 5.22 Drawing a possible enemy, a ninja ready to throw a spear!

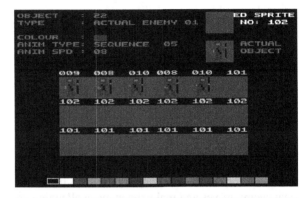

Fig. 5.23 The spear the evil ninja will throw at the player. Enemy bullets have their specific object type. Note also that static objects have a single frame animation (i.e. **Sequence 01**)

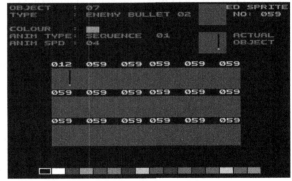

Fig. 5.24 The "**Edit Enemy Bits**" screen where we can define many important parameters related to the enemy. This ninja will throw spears (**Bullet Objects 2**) at any target straight in front of him (**Fire Type** identified by the *down arrow*)

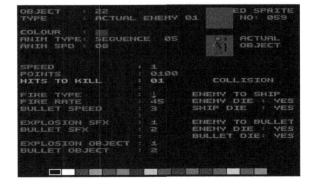

behaviour: here, in fact, we should decide not only which bullet is being used (the spear, in our case, which had an ID of 2) but also the fire rate, the bullet speed, direction and so on.

Once all this has been decided, we can actually proceed to place the ninja on the map and plan for different enemy waves. This is done by selecting the option "**Edit Attack Waves**" from the **Main Menu** (press **A**). A submenu will open asking us if we want to add a new enemy (**F1**), join enemies (**F3**, this would be useful to have sprites move together or for making bigger enemies, like bosses) or delete an existing enemy (**F5**). It will also show how much memory left we have for constructing our waves (i.e. "**Units Free**").

Once pressed **F1** to add an enemy, we will be asked which specific foe we want to add and then we will be able to move around the map via the joystick. This happens at two different speeds: first we can move faster ("**Rough**") and then, by pressing the joystick button, slower ("**Fine**") to have better control on the exact location we want our opponent to appear. Press the fire button again and the enemy will show up, blinking and ready to be placed. Select the starting position you want by moving him around with the joystick and then press the button again. The enemy will stop blinking and any movement we do now will be recorded and replayed in the game once the player reaches that specific map area. Once done, press **F7** to store the whole sequence and go back to the "**Edit Attack Waves**" submenu. Repeat this as many time as you like, memory permitting, to build your own enemy waves and make the game challenging and fun!

Other options available from the **Main Menu** are the "**Edit SFX**", that allows for the definition of sounds to be used for player or enemy fire and explosion effects, and a "**Edit Front End**" screen to arrange a simple splashscreen where, for example, write a brief introduction about the game itself and its creators.

Last but not least, the "**Storage**" option allows for loading and saving the data on either tape or disk, everything or just a specific component like sprites or sound effects, and to finally export the finished game as a standalone program for easy distribution.[9]

---

[9] More information on SEUCK can be found online at http://www.seuckvault.co.uk.

Chapter 6
Windows and Icons

Commodore had very ambitious plans for the C64 from the very beginning, with Jack Tramiel trying to position it in the market as the definitive "Apple II killer". Naturally, to manage this goal, they needed more than games to push the computer's appeal beyond the entertainment field and compete with a machine that had exceptional apps like VisiCalc, the first spreadsheet program to run on home computers, besides a more robust OS.

We already saw, towards the end of Chap. 3, how this was originally thought to be achieved, i.e. by releasing a special cartridge to give the C64 access to the whole CP/M software library. Unfortunately, this plan failed miserably due to incompatibilities that were too much rooted into the system to be changed and Commodore had to innovate by trying a different path, i.e. using graphical user interfaces (GUI), like icons and overlapping windows, to manipulate programs.

The first ever computer lab to experiment with GUIs and icon driven systems was Xerox PARC (standing for "Palo Alto Research Center", based in California) with their Alto computer starting as early as 1973. However, the Alto was pure R&D work and was not meant to be commercialized. Its innovative ideas remained confined mostly within Xerox research facilities till Steve Jobs visited them in the late 1970s and got inspiration for a new, much more intuitive graphic based human-computer interaction paradigm to be implemented first in the Apple Lisa (released on January 19 1983 at the price of $9,995) and then in the Apple Macintosh (released on January 24 1984 for $2,495).

Was there anything in this area that the C64 could do, not only to surpass its direct competitor, the Apple II, but even to get close to cutting edge machines that were obviously much more powerful and costed thousands of dollars more?

Soon after Lisa was released, Commodore managed to publish a conceptually groundbreaking product that, despite all its obvious limitations and shortcomings, actually allowed many computer users to experience an icon driven system to for the first time ever: the "Magic Desk" (Fig. 6.1).

© Springer Science+Business Media Singapore 2015
R. Dillon, *Ready*, DOI 10.1007/978-981-287-341-5_6

Fig. 6.1 The Magic
Desk I : Type and File
(© Commodore 1983).
"Commodore's Answer to
Lisa: Is This The Ultimate
In Friendly Software?" so
titled the magazine "RUN"
(May 1984, p. 18) in an
article introducing the new
cartridge, which was being
sold at $71.95

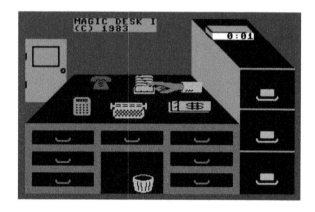

This was a fully graphical program simulating an office environment were icons (here referenced as "Pictorial Commands") could be pointed at and clicked by using a finger-shaped cursor, giving access to a very simple word processor that could format and file documents in the included cabinet, saving them to a floppy disk as well as printing them or deleting them by selecting the corresponding icons. A calculator and a financial planner software were also planned, their icons already appearing on the desk, but a "Magic Desk II" expansion was never released despite the popularity of the first cartridge.

Once again, true to Jacks's original mantra, Commodore tried to make its computers more accessible to the masses by making technology more "user friendly" or, as stated in the marketing campaigns of the time, making computers "people literate" instead of forcing the opposite, i.e. making people "computer literate".

The following years brought many revolutionary changes, both internally with Jack resigning from his position and buying the consumer division of Atari Inc. from Warner Communications, and in the industry at large with the rise of MS-DOS and corresponding decline of CP/M and the release of the Apple Macintosh which was soon followed by the first appearance of Microsoft Windows (1985), showing the world how more convenient a GUI approach was compared to command line based systems.

In this rapidly evolving landscape Commodore and the C64 didn't stay still either. A new GUI based OS able to compete with both MacOS and Windows was commissioned in 1985 to Berkely Softworks, a small startup founded by Brian Dougherty, who previously worked in the game industry at Mattel Electronics and Imagic before the 1983 video game crash wiped out those companies. The "Graphical Environment Operating System" (GEOS) was released in 1986 and showed what the C64 could actually do besides playing games (Fig. 6.2).

Featuring windows, icons, dialog boxes, drop down menus and even slider controls, for example to set up system preferences like mouse speed and acceleration, GEOS was a complex OS that could actually stand its ground against much more expensive competitors.

Files could now be dragged and dropped on the printer icon or into the wastebasket to trigger the respective functions while programs, as well as their respective

Fig. 6.2 GEOS (© Berkeley Softworks 1985). Thanks to being bundled with the redesigned C64c, at its peak GEOS actually became the second most popular GUI based OS by units shipped, after MacOS but ahead of the original Microsoft Windows

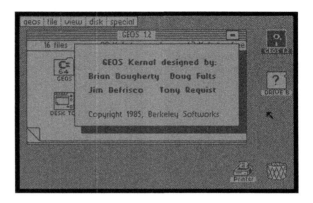

files, could be started automatically by double clicking on them. Different useful programs and utilities were also bundled with the OS, like GeoPaint, GeoWrite (a word processor that allowed for different fonts and formatting options besides offering an impressive print preview feature) and QuantumLink. The latter was an interesting on-line communication program linking to a service offering news, games and an online store that eventually evolved into America Online.

Many different software packages were released, both by Berkeley Softworks and third parties, including GeoCalc, a long due good spreadsheet, GeoPublish, an advanced word processor offering also drawing functionalities, and, despite the memory constraints, even high level programming languages like Basic and Forth.

Later versions of GEOS could handle memory expansions as well as additional hardware and, even today, its legacy lives on in modern software developed by the Breadbox Computer Company.[1]

Perhaps surprisingly, Commodore wasn't the only company pushing for GUI based environments for the C64 and also completely independent third parties got involved. For example, H&P Computers, a Dutch company based in Rotterdam and the producer of the successful multipurpose cart series "The Final Cartridge", enhanced their latest offering with a graphical desktop inspired by the early AmigaOS (Fig. 6.3).

The desktop included a few utility programs like a calculator and a text editor, besides allowing easy access to installed peripherals together with disk and tape fast loaders. Developing new window based programs to run in this environment was actually possible but very few, if any, were ever released. This strongly reduced its overall appeal, with the result that most users perceived it more like an utility suite complementing the cart's hacking features[2] than a fully fledged

---

[1] More information can be found at http://www.breadbox.com/ while a good collection of GEOS resources for the C64, which was released as freeware in 2004, can be found here: http://www.ly onlabs.org/commodore/onrequest/geos.html.

[2] The main cartridge strength was considered to be its robust machine code monitor program which could even allow users to dump and edit text and sprites present in memory.

Fig. 6.3 The Final Cartridge III (© RISKA B.V. Home & Personal Computers 1987). An Amiga inspired desktop environment for the C64!

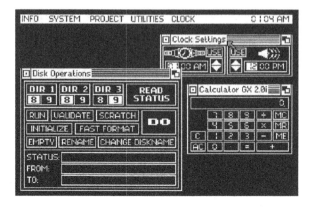

alternative OS. In any case, it was another worthy accomplishment for a computer that made its mere 64 KB of RAM and clock speed of only 1 MHz defy its inherent limitations and always find novel ways to surprise its users.

Chapter 7
BBS: The Internet Can Wait

It may be odd today to think of a world with computers but without the Internet and it may even be more surprising to realize that, even when the Internet as we know it was not available, people still managed to use computers to connect together, socialize and exchange files online. In fact, starting from the late 1970s till the early 1990s, there was something else that allowed users to keep virtually in touch with each other through standard phone lines: Bulletin Board Systems, or BBS. While BBS were mostly a North American phenomenon (accessing phone lines for data transfer required modems, whose penetration in Europe was unfortunately much lower than on the other side of the Atlantic at that time) it is still well worth discussing them to understand the overall impact they had on the computer world, spawning an online subculture of users and geeks, and, ultimately, paving the way for a quick acceptance of the upcoming World Wide Web, which looked to many like a natural evolution of the pre-existing BBS world.

Anyhow, despite the similarities, there were fundamental differences between the BBS and the Internet: contrary to the modern Internet infrastructure where each computer is virtually networked with any other via dedicated servers ultimately making up all the various nodes spanning the whole planet, a BBS generally worked in isolation: it was simply a computer running specific software connected to a phone line that could be accessed from another computer by calling it via a modem.[1] If someone was already connected, the phone line would return a busy signal and the new caller had no choice but to wait and call again later.

Despite this obvious limitation, BBSes spread like wildfire starting from January 1978 when Ward Christensen and Randy Suess wrote the first software to organize a newsletter hosted on their computers that could be accessed externally via a direct modem connection through their phone line. Anyone could call,

---

[1] The first modem (MODulator/DEModulator) was invented at Bell Laboratories in 1963 to transfer data over telephone lines and was then perfected by Dennis Hayes and Dale Heatherington in 1977, enabling these devices to work with different micro-computers.

© Springer Science+Business Media Singapore 2015
R. Dillon, *Ready*, DOI 10.1007/978-981-287-341-5_7

connect and read or write news about the snow storm that just hit Chicago during those days. Most importantly, they also started giving away their newly developed newsletter software for free: within a couple of years, around 300 BBSes were active in the USA alone.

It is important to note that each BBS was its own independent island of information hosting its own community of users. These communities often became closely knit together, with people logging into discuss and share information, upload and download software from the host machine (needless to say, piracy was rampant in those days and games were by far the most downloaded type of programs) or even play some strategy and turn based online games, often inspired by board games like RISK, which, due to the asynchronous nature of the connection, could take a very long time to finish.

All this bursting of new activity over phone lines was a very rewarding experience for this pioneering online community but could also be very expensive due to the inherent costs involved with long distance calls. To keep phone bills under control, users started to cluster around BBSes falling within their own area codes only. While limiting, this wasn't necessarily a bad thing since it helped turning many virtual friendships into real and long lasting ones thanks to gatherings that were often organized on a regular basis.

Equipment, like the modem themselves, was also very expensive, especially in the early days: a 300 baud modem (a speed that allowed to receive just a few characters of text per second) could cost up to a few hundred dollars and, when 1,200 baud modems finally appeared, allowing for much faster file transfers, they could cost up to $1,000. Luckily, things were soon going to change for the better: the first pressure to lower prices happened, once again, when Commodore stepped into make technology more affordable and accessible to a broader range of people thanks to its own line of products.

Already in 1981 Michael Tomczyk managed to have Commodore work on and deliver the first modem under $100, compatible with both the VIC-20 and the upcoming C64. Sold at $99 and packaged with telecomputing services worth almost $200 offered by the likes of CompuServe and Dow Jones, the VICModem (Fig. 7.1) was a resounding success that soon achieved the same sales milestone the VIC-20 did. It was the first modem to sell over 1 million units and can surely be considered responsible for significantly contributing to the expansion and popularity of the BBS world among new home computer users.

The sudden and somewhat unexpected success of this new range of Commodore products even managed to start giving some trouble to their Customer Support centers, which began to receive serious pressure due to a seemingly unlimited flow of calls from new, inexperienced users.

To solve this problem Tomczyk, Michael also arranged for a very innovative approach to community management by relying on CompuServe[2] infrastructure, starting the "Commodore Information Network". This was a dedicated BBS hosted by CompuServe where users could log-in and discuss any technical issue

---

[2] CompuServe was the first major commercial online service provider in the United States.

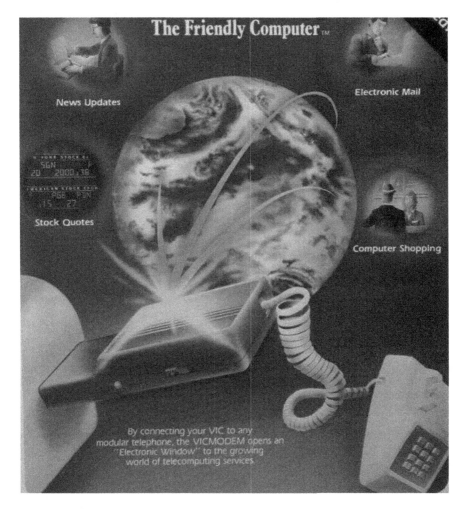

Fig. 7.1 The front of the original VICModem box, showcasing the cartridge connected in between the computer and a telephone. Various services, like stock quotes, news, electronic mail and online shopping were now easily accessible

with each other, ultimately helping each other without the need for official support. The new service was so successful that, in 1982, the network actually accounted for the largest amount of traffic on CompuServe.

In the mid to late 1980s, when the BBS activity was at its peak, it was estimated there were more than 150,000 active ones in North America and, while some required subscription fees and were run as actual businesses, most of them were freely accessible and managed by "System Operators" (sysops) who did all this as an hobby and just for the love of being at the forefront of technology.

As predictable, BBSes had a very sharp decline once the Internet appeared but, even today, a few still exist and some can still be accessed via dialling a phone number even though, for ease of use, most are usually accessed via Telnet.

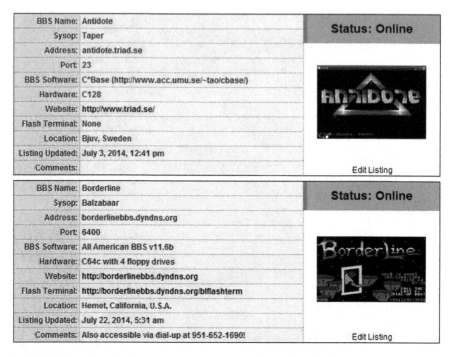

BBS Name:	Antidote
Sysop:	Taper
Address:	antidote.triad.se
Port:	23
BBS Software:	C*Base (http://www.acc.umu.se/~tao/cbase/)
Hardware:	C128
Website:	http://www.triad.se/
Flash Terminal:	None
Location:	Bjuv, Sweden
Listing Updated:	July 3, 2014, 12:41 pm
Comments:	

Status: Online

Edit Listing

BBS Name:	Borderline
Sysop:	Balzabaar
Address:	borderlinebbs.dyndns.org
Port:	6400
BBS Software:	All American BBS v11.6b
Hardware:	C64c with 4 floppy drives
Website:	http://borderlinebbs.dyndns.org
Flash Terminal:	http://borderlinebbs.dyndns.org/blflashterm
Location:	Hemet, California, U.S.A.
Listing Updated:	July 22, 2014, 5:31 am
Comments:	Also accessible via dial-up at 951-652-1690!

Status: Online

Edit Listing

Fig. 7.2 Information about two modern BBSes still running on real Commodore hardware as of July 2014: one hosted in California and one in Sweden, with the former also accessible via dial-up. Screenshot taken from the "Commodore BBS Outpost" website

Specific to the Commodore world, an updated list is currently available from the "Commodore BBS Outpost" website, accessible at http://cbbsoutpost.servebbs.com (Fig. 7.2).

Chapter 8
Verba Volant, Scripta Manent

Latin proverb, literally meaning "spoken words fly away, written words remain".

The computer revolution gave birth also to another related industry: its specialized press. Starting from Byte, whose first issue was published in September 1975, more and more magazines appeared in the following years discussing a broad range of topics, with games being among the most popular and extensively covered on both home computers and gaming consoles.

Interestingly, publishing activities were developed in parallel both by independent groups and by the computer manufacturer themselves who saw this as an effective tool to enhance their marketing strategies while also supporting existing users. By summer 1982, Commodore itself had already released 18 issues of its bimonthly magazine "Commodore MicroComputer", covering mostly productivity topics focusing on the PET, but the huge success of the VIC-20 and the newly released C64 demanded something different. To fill this void, Commodore started a completely new and revised magazine: "Commodore Power/Play" to be published on a quarterly basis. Supervised by marketing director Kit Spencer, it was aimed at emphasizing entertainment and educational uses of the latest computers and was met by a good success till the end of 1986 when it was merged with Microcomputer to become "Commodore Magazine", another publication that remained in print till October 1989.

In any case, the real burst of creativity in the press business happened outside the naturally biased and tight control of the hardware manufacturers themselves and, especially during the early and mid 1980s, many different magazines appeared across North America and Europe to satisfy the thirst for knowledge of the fast growing C64 user base. Clearly, not all were equally successful: some were short lived while others managed to capture the hearts of many readers and are still fondly remembered and collected today.

One of the first specific publications entirely dedicated to Commodore's machines was "Compute!'s Gazette" which had its first issue in July 1983 (Fig. 8.1) and was actually spawned by the already existing "Compute!", a more generic magazine covering different home computers that started as early as 1979.

© Springer Science+Business Media Singapore 2015
R. Dillon, *Ready*, DOI 10.1007/978-981-287-341-5_8

Commodore 64™ **PREMIER ISSUE** VIC-20™

COMPUTE!'S
$2.00
July 1983
Issue 1 Vol. 1, No. 1
63380

GAZETTE

For Owners And Users Of **Commodore VIC-20** And **64** Personal Computers

SKYDIVER
An Exciting Action Game For VIC & 64

Using Joysticks On The Commodore 64

A non-technical explanation of how joysticks work on the Commodore 64, and the use of joysticks in your own programs. Complete with diagrams, examples, and ready-to-use program routines.

An Introduction To BASIC Programming On The VIC-20 & 64

The Beginner's Corner, and other monthly columns and articles, show how programming your own computer can be not only easy, but fun.

COMPUTING FOR KIDS

At last, a monthly column especially for the people who will inherit the computers of the future: the children of today.

Simple Answers To Common Questions
For Beginning Computer Users

Also In This Issue

Does Your Computer Need A Cassette Recorder?

The Programmer Behind *Galactic Blitz*

Liven Up VIC & 64 Programs With Sound

Fig. 8.1 The first issue of "Compute!'s Gazette": one of the longest running publications dedicated to Commodore's 8-bit computers, being published monthly from July 1983 till June 1990. At its peak, the magazine had about 80,000 subscribers

"Compute!'s Gazette" was a comprehensive magazine targeting hobbyists at various levels of experience and featured articles, tutorials and reviews side by side with actual listings to be typed in by users. These were often complex and instructive programs spanning both utilities and games written in Basic or even in machine language, with the magazine also providing listings for utilities aimed at making the machine language input and editing process a little easier.

Type-in programs were a constant also in other successful magazines to appear in North America like "Ahoy!" and "RUN". Both started publication in January 1984 and were relatively similar in scope, featuring high quality programs together with technical articles and reviews of both hardware peripherals and software, spanning productivity and games alike. Ahoy!'s last issue was published in January 1989 while Run managed to survive till the end of 1992. RUN, in its heyday, had a truly astonishing monthly circulation of more than 200,000 copies and was also noteworthy for publishing end of the year special issues that grouped together some the year's best tips, programs and articles together with very useful pullout references for programmers (Fig. 8.2).

The publishing scene in Europe was also very active. With the UK being at the heart of the home computer revolution in that part of the world, it shouldn't be surprising to find out that it was also home to some of the most successful computer magazines of the time.

Among the most well known, "Commodore User" (CU) and "Your Commodore" (YC) had their first official issue in September 1983 and 1984 respectively, with CU being published till 1990, when it changed its name in "CU Amiga" to focus on 16-bit systems,[1] and the second having its last issue in October 1991 instead.

Both magazines were met by good success and mixed a variety of topics, including games, technical articles, type-in listings, programming tutorials and hardware reviews like their American counterparts. However, the growing market for games prompted publishers to gradually expand their coverage year after year, with entertainment ultimately becoming the main focus.

In any case, there was one magazine that perfectly understood the C64 exploding gaming market from the very beginning and didn't have to drastically change its editorial direction later on. That magazine was "Zzap! 64" (Fig. 8.3) and was published by Newsfield Ltd, the same company behind the successful ZX Spectrum magazine "Crash".

With its first issue published in April 1985, Zzap! decided to leave out listings and technical articles to talk exclusively about one hot topic: games. The choice paid off and, arguably, it soon became one of the most successful and beloved C64 magazines of all time, setting the bar for game related journalism for years to come and still fondly remembered by many till this very day.

Part of Zzap! unique appeal was to be found also in its original presentation style: striking cover art[2] inspired by specific games graced each issue while reviewers' comments (each game was discussed by a panel of three reviewers, like a school examination commission) were featured via comic-like drawings, aptly summarizing the value of the game under analysis with facial expressions ranging from enthusiastic to indifferent to disgusted.

---

[1] CU Amiga remained in print till 1998.

[2] Drawn by Oliver Frey.

Fig. 8.2 The first special issue of RUN, published in 1985

Fig. 8.3 Zzap! 64's first cover. We won't find any program to type here, just page after page of great game reviews!

Top notch reviews, with games being rated across different categories like Presentation, Graphics, Sound, Value for Money and so on, by a team of young, talented and passionate journalists were the main obvious strength of the magazine, even if, in hindsight, we can see how they not always appreciated all games as they deserved. For example, groundbreaking titles like "SidMeier's Pirates!" (see Sect. 4.2) or "M.U.L.E.", now regarded as timeless classics, were not fully understood back then. These misunderstandings may actually be seen in retrospect as an additional proof that those games were actually far ahead of their time. Also, the honest and direct tone of the magazine, while definitely an asset to establish the magazine reputation among its readers, did create some friction with developers at times, starting from the very beginning. In the first issue, in fact, a lacklustre review of Jeff Minter's latest game, "Mama Llama", prompted an angry reaction from the dev eloper who addressed the magazine in his own newsletter as "written by 12 year olds for 12 year olds".

Regardless of these few hiccups, Zzap! quickly became the main reference for all C64 gamers and peaked in popularity around 1987 and 1988, as pointed out by the official circulation figures counting more than 80,000 copies sold between July and December 1987. Regrettably, after that, a slow decline started when the editorial direction decided to begin targeting a younger audience instead of growing up along with its existing readers. By 1992 the magazine was something completely different and the then-publisher Euro press decided to re-launch it by turning issue 91 into issue 1 of "Commodore Force", another magazine whose last issue was published in February 1994.

As a further testimony to the enthusiasm and vitality of the C64 community, magazines dedicated to the ageing computer actually kept spawning till the very end. A notable example was "Commodore Format" that, founded by former Zzap! staff, started operations in September 1990 and reached an impressive monthly circulation of more than 60,000 copies in the first half of 1992. The magazine remained in print till October 1995.[3]

However, UK wasn't the only theatre where the computer and gaming press was growing fast and all across Europe the "home computer revolution", with the C64 at its centre, kept spreading and triggering new related ventures including magazines. Germany had dedicated publications such "64'er" and Italy had a localized version of Zzap!, besides a more technical magazine addressing hobbyists and programmers alike named "Commodore Computer Club", which started publication as early as September 1982 with an official endorsement from Commodore itself.

Overall, this was a very lively industry where everyone had a chance of adventuring into new uncharted grounds and, in almost every issue of any magazine, the sense of wonder and discovery was palpable. For the interested reader, such staggering amount of information is luckily being saved for posterity thanks to extensive online archives which are offering scanned versions of these, and many more,

---

[3] http://www.commodoreformatarchive.com/.

old magazines. At the time of writing, websites such as "The Internet Archives"[4] and "DLH's Commodore Archive"[5] are among the best resources for anyone wanting to delve deeper into this topic for historical research or consultation purposes.

---

[4] https://archive.org/details/computermagazines.

[5] http://www.bombjack.org/commodore/magazines.htm.

Chapter 9
Today and Tomorrow

To someone new to the Commodore 64, realising how active its scene has been throughout these years may, perhaps, be even more surprising than discovering its original legacy.

We already had a chance to discuss the SEUCK engine in Chap. 5 and briefly touch on how it is still relevant today, with people actively using it to make new games just for the fun of it and even modifying it to make games with better performance and scrolling in different directions,[1] but the new software and gaming scene has much more to offer than just a SEUCK inspired revival. Besides an original and unique demo scene, which is one of the oldest components of the C64 ecosystem having been producing stunning multimedia demos since the very beginning,[2] there are software houses still producing and selling new games in different formats, both via digital downloads and collectible boxed editions including tapes, floppies or cartridges. Companies like Protovision,[3] Psytronik[4] and RGCD[5] are well known to modern Commodore enthusiasts and every year offer new, high quality releases. Interestingly, these are sometimes also selected by addressing the development community at large via specific events: for example, RGCD has been organizing an annual "16 KB Cartridge Game Development Competition"[6] since 2011, where anyone can submit their C64 games as long as they fit the requirements for a possible 16 KB cartridge edition and, eventually, get chosen for a distribution agreement. Finding new and original ways to keep the hobbyists community engaged, like these game development competitions successfully do, was probably instrumental in keeping the scene itself alive all these years.

---

[1] See note 6 in Sect. 5.5.

[2] Gatherings and events like Datastorm (http://www.datastorm.se/) are still regularly organized.

[3] http://www.protovision-online.com.

[4] http://www.psytronik.net.

[5] http://www.rgcd.co.uk.

[6] http://www.rgcd.co.uk/2014/04/c64-16kb-cartridge-game-development.html.

© Springer Science+Business Media Singapore 2015
R. Dillon, *Ready*, DOI 10.1007/978-981-287-341-5_9

In any case, software isn't the only area of interest and lots of progress has been done also on the hardware side of things with new peripherals able to make the operation of the C64 much easier and convenient. New means for loading old programs have been devised, with several devices like the "SD2IEC"[7] and its derivatives like the "C64SD Infinity" and "Princess"[8] or the "1541 Ultimate II",[9] which are being manufactured to allow the loading of programs from modern memory cards into the original hardware. Very advanced and powerful multipurpose cartridges like the "Turbo Chameleon 64"[10] are also constantly being revised and improved.

Hacking the C64 (with "hacking" used here in the original sense of the word, coined at MIT in the early 1960s, meaning an unconventional or unorthodox application of technology) didn't stop here though, and the C64 can now do completely new things it was never designed for, like surfing the Internet or even posting on Twitter!

This little technological miracle has been achieved only thanks to another modern technological trend that recently gained significant attention: the so called "Internet of things", i.e. allowing even small, underpowered devices and appliances to access the modern internet infrastructure to send or retrieve data, with the idea of enabling countless of different applications, from environmental monitoring to home automation and other industrial uses. To make all this possible, new operating systems able to fit and run on such limited embedded computer devices had to be designed first. One such system is "Contiki",[11] a tiny but fully featured OS including, among other things, a multi-tasking kernel, a windows based GUI, a personal web server and a web browser. Contiki only needs about 30 KB of RAM to work and is currently being used in street lighting systems, radiation monitoring systems and more but, even more importantly for us, its very limited memory footprint makes it suitable to run on old 8-bit computers as well. Being an open source project, it didn't take long to have a port running on our C64 as well[12] (Fig. 9.1).

Unfortunately, the C64 doesn't have an ethernet port and wireless connection is not available either so installing Contiki by itself isn't enough to allow us to surf the internet and special cartridges like the "Retro Replay" and its complementary "Retro Replay Network Card" are also needed, to ultimately enable the C64 to connect to a router and being properly configured. The configuration process is relatively straightforward and, for the curious reader willing to try the excitement of browsing the web on a 30+ year old computer, a very clear tutorial written by Nigel Parker was published on the issue 81 of the downloadable magazine "Commodore Free"[13] (pp. 20–22).

And what can we expect from the future? Where can the scene go from now on? It is clearly impossible to say but the passion shown so far by the C64 community

---

[7] http://www.sd2iec.co.uk/id14.html.

[8] http://www.manosoft.it.

[9] http://www.1541ultimate.net/.

[10] http://www.syntiac.com/chameleon.html.

[11] http://www.contiki-os.org/.

[12] http://csdb.dk/release/?id=14635.

[13] http://commodorefree.com/magazine/vol8/issue81.pdf.

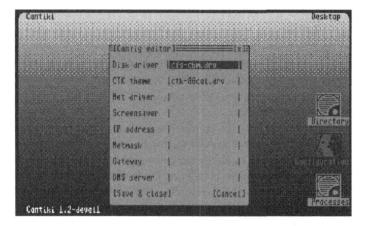

Fig. 9.1 A C64 running Contiki and ready to be configured to surf the internet

is definitely not going to get extinguished anytime soon. Many new ways to enjoy an old but still fascinating way to do computing, playing games or simply "hack" with technology, are surely going to be devised and experimented with.

In this regard, it is interesting to ask the opinion of Mr. Nigel Parker who, in his role of Editor for the "Commodore Free" magazine, has always been closely involved with everything C64 related throughout these years.

Roberto Dillon: Hello Nigel, please tell us a little about how you first discovered Commodore and the C64 back in the day.

Nigel Parker: I was fascinated with computers from an early age: I remember buying magazines on anything computer related and I also had a fascination on electronic music. I had to beg my parents to buy me a VIC-20 having assessed all the other machines and decided this was the best system for my needs: it was colour, had sound and a proper keyboard just like a typewriter. Yet, my parents were unsure and thought it would be another fad so I remember proving to them I would stick at this by typing in the listing for various machines on my mother's typewriter! Not an easy feat with the Commodore listings: for the various commodore characters I would just type what I saw, so the clear screen heart GFX, for example, would by just typed {heart}. I had no idea what they meant anyway!

Later my father took me to a friend of his who had just purchased a VIC-20. Straight away I dived in and showed them how to load games and applications, as our friend seemed clueless. This, along with the broad range of educational titles readily available, convinced my mother and farther that a VIC-20 was indeed a good investment and so, at Christmas, under the tree was a VIC-20! I spent all Christmas trying to get it to print HAPPY CHRISTMAS and play a carol or "Old Lang Sine". I did succeed although it was rather robotic!

The Vic helped at school too: with me being shy, I couldn't relate to people very well, but with other people talking about computers I could chat all day about games and our really bad BASIC programming skills.

The Vic soon gave way to the Commodore 64 as a natural progression and with the SID chip… WOW! You had a sound synthesiser right in your hands! Hardware sprites gave an arcade quality to games and later came a disk drive and I bought a monitor with saved pocket money! I also bought a printer and Geos which I still use today with an upgrade called Wheels: it can even output to postscript printers and I could theoretically produce the magazine with this setup, however the time restraints of real life prevent this.

RD: Even after the demise of Commodore, you, like many others, still decided to treasure these old computers instead of just forgetting about them and moving along. Do you think there is anything particular that kept your love for the C64 in particular, as well as Commodore at large, alive all these years?

NP: I still have my original Vic and C64 while I sold my Amiga 500 to help paying for an Amiga 4000. What do I love? I love the feelings of playing sid music, I love the tunes, I love the loading graphics and the suspense of the game starting, I love the simplicity of the games, picking up the joystick, waggle it and press fire! Not sure If I loved Commodore as such, mainly because some of the machines they later released were all promise and no delivery. And of course some of the missed machines that never appeared, like the C65, were huge disappointments. Maybe I was a Jack Trammel fan, maybe it was the love the designers put into the machines, maybe it was something perfect in design. Heck, I can sit down now and load a game like "Monty On the Run" and instantly I am taken back in time when I was sitting with my friend in my bedroom, after walking back from town with the game and placing it in the drive and typing load "*", 8, 1. The music was amazing: I remember we sat listening to it for hours, I even recorded the music on an audio cassette and we played it all the time! Sometimes, especially with things like GEOS, it's the thrill of actually getting all your setup working!

Psychologically I suppose it sends me back to a happy childhood, no doubt someone has a technical word for it, it was and still is happy days for my computing. What I didn't know for some time was that the internet had thousands of people like myself. I think it was in 1998 that I had a decent enough connection to click a search page and type "Commodore 64". Imagine how many pages were C64 related … THOUSANDS! I made some good friends this way, searching for other enthusiasts and then emailing them. In some ways, it's like I am still a child, and with new games being released and people reworking the hardware with extra add-ons like SD card readers and network cards and even software that many though would be impossible to create, I think the Commodore name will still live for some time. Nonetheless, I don't think it's just childhood memories: as I once said, a good game is a good game. It doesn't matter too much about the hardware. If a game is ported to the C64, like some of the recent flash games, then you feel like it's an honour to play this as the designer have worked so hard to cram the game onto the system.

RD: You are the Editor of 'Commodore Free', a well known multiformat online magazine completely free to download. This is obviously a work of love that you have been carrying along for several years by now. What keeps motivating you in this endeavour? How and where do you find all the contents to write about in each issue?

NP: Well, it is multiformat in that it covers all Commodore machines, not just the C64. When I started, it was because the magazine "Commodore Scene" had closed its doors and now I needed a magazine I could read in the bath! So I collated some information and released issue 1 mainly for myself! It went mad with hundreds of downloads! By issue 5, my Internet Service Provider said I had to upgrade my package as the traffic was so high I was averaging two thousand downloads per issue! It is not been easy as I am not a "real" Editor. It is all a labour of love but I am well aware it is no way professional: spelling mistakes and poor grammar make up most of the text. However, the enthusiasm I put into each issue I hope shows through. I don't make money from the magazine at all and am luck to break even from donations mainly. Still, to have someone say they read it and liked it is a nice feeling. I also wanted a real D64 or disk version you could read on a real machine as I know some users who won't use a pc, only the C64 and would dial into a BBS to grab a copy the sysop had put on line! Yes, some people don't have internet access apart from dial up and still can access the magazine! Then again some people print out the magazine and give copies away at meetings for free to users without even dialup access. To some it is their only source of information about Commodore. When someone like Bil Herd[14] writes to you and says he enjoys reading the magazine you know you are doing something right!

RD: Is there anything particularly striking and impressive that was developed or accomplished by the C64 scene during these last few years that you think every old and new C64 enthusiast should know about and be proud of?

NP: That's an unfair question, really, because I can't give a one thing answer! Some of the developments like SD card readers and network cards are keeping the machines alive, giving the machines new breath. I would also mention hardware like the DC2N,[15] a way of storing tape files on an SD card so it loads like a tape but, of course, is a digital file. I looked for something like this for years but was told it was impossible. Then I was contacted by Luigi Di Fraia, the device creator himself, and bought one. Absolutely amazing! We also have things like the "Retro Replay" card, that adds features like those in the old "Action Replay" carts but with a network option, and of course the "1541 Ultimate".[16] Software that would have been considered to be impossible back in the day is now readily available and screen modes that didn't even exist back then, for example extending the number of visible colors on the Commodore 128, have been implemented thanks to very smart tricks. I could go on all day, but would still miss out something really important! I know someone will be screaming now about me missing their hardware! For example the "64HDD",[17] a way to use an old laptop as a drive for the Commodore machines. Then we can't dismiss emulators either as some new

---

[14] Bil Herd was the computer engineer who designed the Plus/4, C16 and C128 while working at Commodore in the mid 1980s.

[15] http://www.luigidifraia.com/c64/dc2n/.

[16] http://www.1541ultimate.net/content/index.php.

[17] http://www.64hdd.com/.

users come to the scene through an emulator first and only later may decide to look for a real machine. Also, without emulation some of the new games and demos wouldn't have been created. We can freeze the machine and change settings then restart it. Emulation is also good for running real software hundreds of times faster than would have been back in the day. If you use an assembler on an emulator, it compiles in seconds!

RD: Do you think such old hardware and software, like the C64, its programs and games, still have something valuable to teach new generations of enthusiasts and game developers who don't have, like us, nostalgic memories to fuel their interest?

NP: I think people are still lacking "a way into" computing. Sure, we have things like the "Raspberry Pi" now but nothing is like the old 8 bit machines. If we power on a C64, the machine instantly comes back READY! And at that point we can instantly use it, either by typing direct commands or indirectly by loading a software application or program. Also, I have met a number of people who weren't even born when the C64 was in its "hey days" and yet they love the basics of the machine and how instant things are. Simplistic, yes, but also easy to get to grips with. 8 bit machines are a perfect way into the computing world. It's not just nostalgia, as recent hardware developments have taught people a craft they can then develop into a full time job. People have also taught themselves the art of designing hardware and having devices mass produced, then moved on to bigger jobs and companies as these can see the potential with their newly acquired engineering skills.

Last but not least, we should also remember it was the original designers together with the vision of people like Jack Tramiel that made Commodore what it was. Once they left, the company was just another business and soon its magic faded. I still think companies need to learn this lesson: it's the people and products that make the company, not the directors!

RD: Any forecast for the C64 future in the years to come? Do you think the scene will remain as active as it is today?

NP: I can see the scene carrying on and, yes, "young blood" is entering the scene. This is a very good thing. We still have the so called "elite" users who seem to look down on everyone who can't code in assembler to a very high standard but, for each of these, there are possibly twenty other users eager to help out newcomers and beginners. The main problem is how long the real machines will last. The technology is degrading over time and spare parts are getting rare and expensive. It would be a shame if this would be reduced to just an emulator scene. Emulators do play an important part as they make development times shorter and easier to test, as developers can freeze the machine and look at all its memory on an emulator, but keeping the real machines alive is still very important to get the full experience!

If anything, things seem to be getting busier in the scene: once, where a meeting had two or three people only, now there are ten or twenty. The UK seems slower than other countries in this rediscovery but I guess we are more reserved. The problem for most people is the lack of time. If only I had paid attention in school, I would have had more time to play with the real machine, Now work commitments mean I only get to play a couple of times per week, if I am lucky.

The really unique thing about the C64 is that you turn it on and it says "READY"! A friend back in school once said "ready for what?" Quick as a flash, I replied "READY! For anything you want to use it for!" Even today, I have a word processor on a cartridge and, with this, I can "BOOTUP" quicker than any PC and start typing a letter. I actually did a bank letter on my C64 and had finished and printed it before my pc had even booted to a logon window!

Appendix A
Collecting for the Commodore 64

With more than 10,000 games and applications officially released back in the day, the Commodore 64 is surely a challenging, though potentially very rewarding, system to collect for. Before embarking in such a quest, some clear goals and guidelines should be established first.

First of all, collecting games for the C64 has some fundamental differences compared to collecting for other classic systems like the Atari 2600, the Intellivision or the NES. While obscure games can be very valuable due to their rarity, regardless of their quality, on dedicated systems which had only a few hundred games available (Air Raid, which is valued at more than $30,000 for the Atari 2600, is a typical example), on the C64, with thousands and thousands of games, not many people really care for a low quality, forgotten title that was released unsuccessfully in limited quantities, whatever its supposed "rarity" rating. On the other hand, most C64 collectors tend to focus on games that were high quality, rare box variations and first or special editions of famous games.

Why is it so? Most likely, piracy is one of the reasons that made these games highly collectible nowadays. Despite being considered "popular" back in the day, this didn't mean most kids who played and loved them actually owned a legit copy. Now that many of them turned into more responsible adults and collectors, effectively owning the "real" thing becomes a common collecting goal that makes such releases very sought after.

It should also be pointed out that, while monetary values can fluctuate wildly from time to time and it is very difficult to provide reliable estimates, lose games have very little value and only complete specimen with box, manual and any eventual extras in good condition, are worth any money as collectible items.

In general, though, we can say that most, if not all, original first releases of the games discussed throughout the book do have at least some collectible value. Well known classics like M.U.L.E., Maniac Mansion, Zak McKraken and the Alien Mindbenders (Lucasfilm 1988), Wasteland (EA 1988) and games by legendary designers like Richard Garriott (aka Lord British: his UltimaI and Ultima III, the

© Springer Science+Business Media Singapore 2015
R. Dillon, *Ready*, DOI 10.1007/978-981-287-341-5

Fig. A.1 The C64 floppy disk version of Chuck Rock is extremely rare and can fetch very high prices. A mint specimen was sold on ebay in 2013 for 1,037 Euros

first in the series to feature a party of adventurers instead of a single character, are particularly sought after) tend to be highly collectible and, in online auctions, can often pass the $100 mark.

Also of great interest to collectors are high quality games that had short but eventful lives on store shelves. For example Great Giana Sisters, a game that had to be withdrawn soon after release due to legal issues raised by Nintendo for Giana's supposed similarities with Super Mario Bros., or high profile games that were announced and then never officially released like Gauntlet III or Chuck Rock, a port of a well known Amiga game that, after being reviewed by Zzap!with an astounding 96 %, was distributed only in very limited quantities only in Italy, tend to be extremely valuable (Fig. A.1).

Box variations of well known games can also trigger collectors' interest. For example, a specific box with alternative art for Commodore's is another item that can demand very high prices (Fig. A.2).

When collecting, it should also be noted that magnetic media like cassettes and floppy disks have a limited life span and, soon or later, data will become unreadable. They will still remain artefacts of the past worth preserving but games won't load any longer. This is also one of the reasons why games and software in cartridge format are usually considered the most collectibles, followed by floppy disks and cassettes respectively. Still, collecting historically important games on floppies can be a very fulfilling experience and there are quite a few games whose first editions are worth high evaluations when in good cosmetic condition and complete.

Besides the already mentioned Great Giana Sisters, Chuck Rock and Gauntlet III, other very rare floppy games to track down are Bounty Bob Strikes Back, Congo Bongo, Katakis and Moonfall (21st Century Entertainment 1991), while, at the moment of writing, the top five rarest collectible games on floppy disk listed on the Rarity Guide website,[1] are:

---

[1] http://www.rarityguide.com/c64_view.php?SortDirection=desc&SortField=6&recordsPerPage=100.

Fig. A.2 The highly
collectible Wizard of Wor
box variation

1. Castles of Doctor Creep
2. Impossible Mission
3. Ultima III: Exodus
4. Sword of Fargoal
5. M.U.L.E.

Tape based games, being the most popular distribution format by far back in the day, are usually not as collectible even though some specific releases, like the Mercenary Compendium Edition, are notable exceptions (Fig. A.3).

Regarding cartridges, a few hundred games and utilities were officially released and the following is a reference table in alphabetical order for more than 230 selected cartridges rated according to rarity and historical importance. Rarity ratings here have been extrapolated from Matt Allen's excellent Mayhem website,[2] among others, as well as from the author's own research, tracking cart based games on online auction sites like ebay and marketplaces on websites such as Lemon64.[3]

Rarity Legend:

1–2: Very Common
3–4: Common

---

[2] http://www.mayhem64.co.uk/.

[3] http://www.lemon64.com/forum/.

Fig. A.3 Mercenary
Compendium Edition,
including the original
Mercenary game plus the
expansion "Mercenary: The
Second City", is one of the
few tape based games that are
very much sought after by
collectors

5–6: Uncommon

7–8: Scarce

9: Rare

10: Extremely rare

Historical Importance is rated as Low (L), Medium (M) and High (H) and is determined
by evaluating the game significance, influence and critical reception (Table A.1).

Table A.1 C64 Cartridge Rarity Table

Game	Year	Company	Rarity	Historical importance	Notes
A Bee Cs	1983	Commodore	6	L	Educational game. Needs magic voice module
Action Replay MK VI	1990	Datel Electronics	4	H	The last and most well known of the Action Replay carts for the C64
Adventure Creator	1984	Spinnaker	7	M	The only game engine for the C64 released in cartridge format
Aegean Voyage	1984	Spinnaker	8	L	
After the War	1990	Dinamic	10	L	
Alf in the Color Caves	1984	Spinnaker	4	L	

(continued)

Table A.1 (continued)

Game	Year	Company	Rarity	Historical importance	Notes
Alien Sidestep	1983	Mr. Computer	8	L	Space Invaders variation
Alpha Build	1984	Fisher Price	3	L	Educational game
Alphabet Zoo	1983	Spinnaker	3	L	Educational game
Amazing Maze	1983	Mr. Computer	7	L	
Arcade Classic Pack	1984	HES Australia	7	L	4 games compilation featuring Frogger, Beamrider, Decathlon and River Raid
Arnie Armchair's Howzat	1984	Armchair Entertainment	9	L	Cricket game
Aspar GP Master	1989	Dinamic	10	L	
Astroblitz	1983	Creative	5	L	Defender clone
Astromarine Corps	1990	Dinamic	10	L	
Avenger	1983	Commodore	3	L	Space invaders clone
BadLands	1990	Domark	7	L	Super sprint clone
Batman	1990	Ocean	4	M	
Battle Command	1991	Ocean	6	H	One the most impressive technical achievements on the C64
Battlezone	1983	Atarisoft	6	H	Conversion of the classic Atari arcade game
BC's Quest for Rires	1983	Sierra on Line	6	H	Generally considered as the grandfather of modern endless runner games
Beamrider	1983	Activision	4	M	
Big Bird's Fun House	1984	CBS Software	4	L	
Big Bird's Special Delivery	1984	CBS Software	5	L	
Block Hopper	1983	Fantasy Software	10	L	Q*Bert clone
Blueprint	1983	Commodore	4	L	Conversion of an obscure Bally Midway arcade game
Bridge 64	1983	Handic	8	L	
Bubble Burst	1984	Spinnaker	5	L	
Bubble Busters	1984	Maxion	7	L	Alternative release of bubble burst
Buck Rogers	1983	Sega	4	L	
Bug Crusher	1983	Mr. Computer	8	L	Frogger meets PacMan
C64GS Compilation Cart	1990	Commodore	4	L	Included with the C64GS console. Includes International Soccer, Flimbo's quest, Klax and Fiendish Freddy

(continued)

Table A.1 (continued)

Game	Year	Company	Rarity	Historical importance	Notes
Calc Result Easy	1983	Handic	6	M	A spreadsheet program
Castle Hassle	1983	Roklan	7	L	Clearly inspired by Exidy's arcade game venture
Centipede	1983	Atarisoft	5	H	Conversion of the popular Atari arcade
Chase HQ 2	1990	Ocean	3	M	
Checkers	1984	Yu Can Software	9	L	
Choplifter!	1982	Broderbund	6	H	One of first C64 cart games to be released
Close Encounters of the Worst Kind	1983	Mr. Computer	9	L	Space invaders clone
Clowns	1982	Commodore	1	M	Requires paddles
Coconotes	1984	CBS Software	7	L	
Congo Bongo	1983	Sega	5	M	Only 2 levels converted from the arcade game (the disc version has all 4 instead)
Cosmic Combat	1983	Maxion	7	L	Alternative release of Spinnaker's cosmic life
Cosmic Life	1983	Spinnaker	6	M	One of the most interesting games in the Spinnaker catalog, based on the well known Conway's "game of life" cellular automaton concept
Crisis Mountain	1983	Creative	6	L	
Cup Final 64	1983	Handic/ Commodore	8	M	Swedish release of Commodore's classic International Soccer
Cyberball	1990	Domark	8	L	Conversion from a Tengen arcade game
Dance Fantasy	1984	Fisher Price	5	L	A game about choreographing dance routines
Dancing Feats	1983	Romox	9	L	Music creation tool
Decathlon	1984	Activision	5	H	A classic in the sports genre
Defender	1983	Atarisoft	5	H	Conversion of the classic Williams arcade
Delta Drawing	1983	Spinnaker	5	L	A drawing program aimed at kids
Designer's Pencil	1984	Activision	6	M	A very nice graphic design package by Gary Kitchen

(continued)

Table A.1 (continued)

Game	Year	Company	Rarity	Historical impor- tance	Notes
Diamond Mine	1983	Roklan	8	L	
Dig Dug	1983	Atarisoft	5	H	Good conversion of Atari's popular arcade game
Donkey Kong	1983	Atarisoft	6	H	Avrage conversion of Nintendo's classic
Dot Gobbler	1983	Mr. Computer	7	L	Pac Man clone
Double Dragon	1989	Melbourne House	9	L	
Double Dragon	1992	Ocean	10	L	Ocean's last C64 cart., made available only at a computer show
Dragon's Den	1983	Commodore	6	H	Similar to Lazarian. By Commodore star pro-grammer Andy Finkel
Ducks Ahoy!	1984	CBS Software	4	L	
Ernie's Magic Shapes	1984	CBS Software	3	L	Educational game
Espial	1984	Tigervision	7	L	Terra Cresta clone
Facemaker	1983	Spinnaker	2	L	
Falconian Invaders	1983	Turbo Software Inc.	9	L	Similar to Buck Rogers
Final Chesscard	1989	TASC	8	M	One of the best chess programs for the C64
Final Cartridge, The	1985	H&P Computers	7	M	Programmer/hacker cart
Final Cartridge II, The	1986	H&P Computers	8	M	Programmer/hacker cart
Final Cartridge III, The	1987	H&P Computers	5	H	Programmer/hacker cart plus a GUI based OS
Financial Advisor	1983	Commodore	1	M	
Fourth Sarcophagus, the	1983	Handic	8	L	Text adventure
Fraction Fever	1983	Spinnaker	4	L	Educational game
Frog Master	1983	Commodore	2	L	
Frogger	1983	Parker Brothers	4	H	Conversion of the classic arcade
Frogger 2	1984	Parke Brothers	6	M	
Funplay Cartridge	1990	Codemasters	7	L	Collection including Pro Skateboard, Pro Tennis, Fast Food
Galaxian	1984	Atarisoft	5	H	Conversion of the classic arcade
Galaxions/Munchman	1983	HES Australia	8	L	Galaxian and PacMan clones
Gateway to Apshai	1983	Epyx	6	H	

(continued)

Table A.1 (continued)

Game	Year	Company	Rarity	Historical impor- tance	Notes
Ghostbusters	1984	Activision	7	H	
Gold Record Race	1984	Maxion	7	L	Alternative release of Spinnaker's Jukebox
Gorf	1983	Commodore	3	H	Conversion of the classic arcade
Graf 64	1984	Handic	6	M	A drawing/plot software
Gridrunner	1982	HES	5	H	Jeff Minter's prequel to Matrix
Gridrunner II: Attack of the Mutant Camels	1983	HES	4	M	American release of the game Matrix
Gridrunner 64	1982	HES	8	H	Swedish release of Gridrunner
Gyruss	1984	Parker Brothers	6	H	Excellent conversion of the classic arcade game
Halftime Battlin' Bands	1984	CBS Software	6	L	
HERO	1984	Activision	6	M	Classic cave exploration game
HESMon 64	1983	HES	4	M	Classic machine lan- guage monitor
HES Writer 64	1982	HES	5	H	A good word processor in only 8k
Hop Along Counting	1984	Fisher Price	3	L	Educational game
International Soccer	1983	Commodore	1	H	The best football/soccer game of its time
Jack Attack	1983	Commodore	3	H	A Commodore classic, inspired by Jack Tramiel and his aggressive way of scolding employees
James Bond	1984	Parker Brothers	7	M	
Jawbreaker	1983	Sierra OnLine	8	L	A Pacman inspired game
Juice	1983	Tronix	9	L	A Q*Bert inspired game
Jukebox	1984	Spinnaker	5	L	
Jumpman Junior	1983	Epyx	6	H	
Jungle Hunt	1983	Atarisoft	5	M	
Jupiter Lander	1982	Commodore	2	H	First ever C64 game, bundled with the C64 itself in August 82. Developed by HAL Laboratories after the original 1979 Atari arcade game

(continued)

Table A.1 (continued)

Game	Year	Company	Rarity	Historical importance	Notes
Kickman	1982	Commodore	3	H	Conversion of Bally Midway arcade and the second ever C64 game
Kickman 64	1982	Handic	8	H	Swedish release of Kickman
Kids on Keys	1983	Spinnaker	2	L	Educational game
Kindercomp	1983	Spinnaker	3	L	Educational game
Kung Fu Master	1986	US Gold	7	M	Conversion of the classic arcade
Laser Cycles	1983	Turbo Software Inc.	7	L	Based on Tron's laser cycle scene
Last Ninja Remix	1990	System 3	8	H	Last Ninja 2 with new graphics and sounds
Lazarian	1983	Commodore	2	L	Conversion of the arcade game. By Andy Finkel
Laser Zone	1983	HES	7	M	By Jeff Minter
Leaderboard	1987	Access	7	L	
Le Mans	1982	Commodore	3	M	By HAL Laboratories. Requires paddles
Learning with Leeper	1983	Sierra Online	3	L	Educational game
Letter Go Round	1984	CBS Software	3	L	Educational game
Letter Scrambler	1983	Maxion	7	L	Alternative release of Spinnaker Up for Grabs
Linking Logic	1984	Fisher Price	3	L	
Lode Runner	1983	Broderbund	6	H	One of the very st games to include a level editor
Logic Levels	1984	Fisher Price	4	L	
Lunar Leeper	1983	Sierra Online	8	L	
Magic Desk I	1983	Commodore	4	H	One of the first icon driven environments for home computers
Make a Face	1983	Learning Tree	8	L	Licensed from Spinnaker Facemaker
Mario's Brewery	1983	Mr. Computer	8	L	Donkey Kong clone
Mastertype	1983	Scarborough	5	L	Educational game
Math Mileage	1984	CBS Software	6	L	Educational game
Maze Man	1983	Turbo Software Inc.	5	L	PacMan clone
Maze Master	1983	HES	6	M	By Michael Cranford, before he designed the Bard's Tale
Memory Manor	1984	Fisher Price	3	L	

(continued)

Table A.1 (continued)

Game	Year	Company	Rarity	Historical importance	Notes
Miner 2049er	1983	Big 5	7	M	Challenging early platformer
Minnesota Fat's Pool Challenge	1983	HES	5	L	Known as Hustler in the UK
Monster Voyage	1984	Maxion	8	L	Licensed from Spinnaker Aegean Voyage
Moon Patrol	1983	Atarisoft	5	H	Excellent conversion of the classic arcade
Moondust	1983	Creative	6	L	
Motor Mania	1982	Romox	9	L	
Mountain King	1983	Beyond	6	L	
Movie Musical Madness	1984	CBS Software	6	M	Interesting attempt at having a musical movie maker. For kids
Mr Cool	1983	Sierra Online	7	L	Q*Bert clone
Mr TNT	1984	HES	7	L	
Ms PacMan	1983	Atarisoft	5	H	Conversion of the classic arcade
Muistio 64	1984	Handic	9	L	Word processor
Music Composer	1982	Commodore	1	M	Early music composition tool. By Andy Finkel
Music Machine	1982	Commodore	1	L	
Mutant Spiders, the	1983	Handic	8	L	Text adventure
Myth	1990	System 3	9	H	Generally considered as one of the best games for the C64
Narco Police	1990	Dinamic	10	L	
Navy Seals	1990	Ocean	6	H	Tie in with the movie of the same name
Nova Blast	1984	Imagic	8	M	A port showing how Imagic tried to convert itscatalog after the 1983 Crash
Number nabber shape grabber	1983	Commodore	3	L	Educational game
Number Tumblers	1984	Fisher Price	4	L	Educational game
Oil's well	1983	Sierra Online	6	L	Clone of Stern Anteater arcade
Omega Race	1982	Commodore	2	H	Conversion of Bally Midway arcade
Omega Race 64	1982	Handic/ Commodore	4	H	Swedish release

(continued)

Table A.1 (continued)

Game	Year	Company	Rarity	Historical impor- tance	Notes
Pac Man	1983	Atarisoft	5	H	Conversion of the iconic game. No intermission scenes
Paint Brush	1983	HES	5	M	Drawing program
Pancho	1984	Romox	9	L	Q*Bert clone
Pang	1990	Ocean	6	M	Convertion of Mitchell's 1989 arcade. Probably inspired by a 1984 Zx Spectrum game named "Bubble Buster" by Hudson Soft
Park Patrol	1984	Activision	7	M	
Pastfinder	1984	Activision	8	H	
Peanut Butter Panic	1984	CBS Software	4	L	Educational game to teach kids cooperation and wise use of resources
Pinball Spectacular	1983	Commodore	4	M	By HAL Laboratories
Pipes	1983	Creative	6	L	
Pit, the	1983	HES	8	H	A little known but seminal game that inspired Boulder Dash. Conversion of the 1981 Centuri arcade
Pitfall	1984	Activision	6	H	The Atari VCS master- piece on the C64
Pitfall 2	1984	Activision	6	M	
Pitstop	1983	Epyx	4	H	
Pole Position	1983	Atarisoft	5	H	Port of the classic arcade
Popeye	1983	Parker Brothers	6	H	Port of the classic arcade
Powerplay cartridge	1990	Microprose	8	M	Compilation includ- ing Rick Dangerous, Stunt CarRacer and Microprose Soccer
Princess and the Frog	1983	Romox	9	L	Frogger clone in a fan- tasy setting
Q*Bert	1983	Parker Brethers	5	H	Excellent port of the arcade classic
Rack'em Up	1983	Roklan	6	L	Pool game
Radar Rat Race	1982	Commodore	3	H	Rally X clone
Ranch	1984	Spinnaker	6	L	Build your own ranch
Retro Ball	1982	HES	4	L	
River Raid	1984	Activision	5	H	Port of the classic Atari VCS game

(continued)

Table A.1 (continued)

Game	Year	Company	Rarity	Historical impor- tance	Notes
Robocop 2	1990	Ocean	4	M	
Robocop 3	1992	Ocean	5	M	
Robotron 2084	1982	Atarisoft	6	H	Port of the classic arcade. Includes a 2 joystick options
Rootin 'Tootin'	1983	HES	6	M	Port of the Data East arcade. Inspired Taskset game Jammin'
Sammy Lightfoot	1983	Sierra Online	8	L	
Satan	1990	Dinamic	10	L	
Save New York	1983	Creative	6	L	
Seafox	1982	Broderbund	7	L	Sea Wolf clone
Seahorse Hide n Seek	1984	CBS Software	5	L	
Sea Speller	1984	Fisher price	4	L	Educational game
SeaWolf	1982	Commodore	2	H	Port of the classic arcade from Bally Midway
Serpentine	1982	Broderbund	6	L	Inspired by Konami's Jungler (1981)
Shadow of the Beast	1990	Ocean	5	H	Technically impressive port of the classic Amiga game
Simons' Basic	1983	Commodore	3	H	Adds 114 commands to Basic 2.0
Skaermtrolden Hugo	1990	SilverRock	9	L	
Solar Fox	1983	Commodore	4	M	Port of Bally Midway aracde
Song Maker	1985	Fisher Price	7	L	
Space Action	1983	Handic	9	L	
Space Gun	1992	Ocean	6	L	
Space Ric-o-shay	1983	Mr. Computer	7	L	
Space Shuttle	1984	Activision	6	M	
Speed/Bingo Math	1982	Commodore	3	L	
Spitball	1984	Creative	4	L	
Spy Hunter	1984	Sega	8	H	Conversion of the arcade game
Star Post	1983	Commodore	3	L	Tempest variation
Star Ranger	1983	Commodore	4	L	
Star Trek	1983	Sega	5	H	Conversion of the arcade game
Star Wars	1983	Parker Brothers	6	H	Sprite, not vector, based conversion of the arcade game

(continued)

Table A.1 (continued)

Game	Year	Company	Rarity	Historical importance	Notes
Stat	1983	Handic	6	L	A stat, graph and probability tool
Stix	1983	Supersoft	9	L	Qix clone
Story Machine	1983	Spinnaker	3	L	A simple experiment in interactive storytelling for kids
Super Expander	1983	Commodore	5	M	Basic extension similar to Simons Basic, with less but more focused commands
Super Games Cartridge	1988	Commodore	5	L	German release including International Soccer, Colossus Chess 2.0 and Silicon Syborgs
Super Smash	1983	Commodore	4	L	Breakout clone
Super Zaxxon	1984	Sega	7	M	Different game from HES tape/floppy release
Tank Wars	1983	Mr. Computer	8	L	A variation of the classic Atari Tank/Combat games
Tapper	1984	Sega	3	M	Conversion of the arcade game
Tennis	1984	Imagic	7	L	Also known as Matchpoint Tennis
Terminator 2	1991	Ocean	4	M	Movie tie-in
Text 64	1983	Handic	9	L	Word processing
Threshold	1983	Sierra Online	5	L	Space Invaders variation
Timebound	1984	CBS Software	5	L	Educational game
Toki	1991	Ocean	4	L	Conversion of the arcade game
Tooth Invaders	1982	Commodore	2	M	Educational game
Toy Bizarre	1984	Activision	5	M	Good clone of Mario Bros without the… bros!
Trashman	1983	Creative	6	L	Pac man clone
Turtle Graphics 2	1983	HES	4	M	Program your own drawing patterns
Tyler's Dungeon	1983	Turbo Software Inc.	9	L	A prize of $1,000 was offered to whoever finsihed the game first
Ultrex Quadro Maze	1983	Turbo Software Inc.	9	L	
Up 'n Add 'Em	1984	Fisher Price	3	L	Educational game
Up 'n Down	1984	Sega	5	M	Conversion of the arcade game. A classic on many platforms

(continued)

Table A.1 (continued)

Game	Year	Company	Rarity	Historical impor- tance	Notes
Up for Grabs	1983	Spinnaker	4	L	Sort of Scrabble
Viduzzles	1983	Commodore	5	L	Jigsaw puzzle game
Viking Raider	1984	Interphase	8	M	An odd Cart + Floppy combo
Vindicators	1990	Domark	7	L	Conversion of the arcade game
Visible Solar System	1982	Commodre	1	H	The birth of edutainment software
Webster: The Word Game	1983	CBS Software	5	L	Educational game
Wizard of Id's Wiztype	1984	Sierra Online	7	L	Typing game
Wizard of Wor	1982	Commodore	5	H	Conversion of the arcade game
Wonderboy	1987	Activision	8	M	Conversion of the arcade game
Zaxxon	1984	Sega	6	M	Different from the more common Synapse Software tape/floppy version
Zenji	1984	Activision	5	M	
Zone Ranger	1984	Activision	6	M	Conversion of the Atari VCS game, based on Asteroids and Sinistar

Regarding cart games and programs it should be noted that, even for those considered common to find lose, they may actually be not so easy to find complete and in good condition and mint specimen can still sell for relatively high amounts. Also, any Japanese release of Commodore games, identified by the "Max" label, while not explicitly listed here, are all considered to be around rarity 9 and very collectible.

Appendix B
Online Resources

Emulators:

- VICE (Win, MacOS and more, Free): http://sourceforge.net/projects/vice-emu/
- Frodo (Win, MacOS and more, Free): http://frodo.cebix.net/
- C64 Forever (Win): http://www.c64forever.com/
- C64.Emu (Android): https://play.google.com/store/apps/details?id=com.explusalpha.C64Emu&hl=en
- Hand BASIC (iOS, Free): https://itunes.apple.com/en/app/hand-basic-cbm-flavor/id394924289

Development Environments and Tools:

- C64Studio: http://www.nightfallcrew.com/11/12/2011/c64-studio-v1-9/
- CBM prg Studio: http://www.ajordison.co.uk
- CC65: http://www.cc65.org/
- Kick Assembler: http://theweb.dk/KickAssembler/Main.php
- Relaunch64: http://popelganda.de/relaunch64.html
- Sprite Pad and Char Pad: http://www.coder.myby.co.uk/spritepad.htm, http://www.coder.myby.co.uk/charpad.htm
- Sprite Charset Ripper: http://www.studiox64.co.uk/c64ripr.php

Online Magazines:

- Commodore Free: http://commodorefree.com
- Reset: http://reset.cbm8bit.com
- Scene World: http://sceneworld.org
- Lotek64 (in German): http://www.lotek64.com
- Commodore Fan Gazette (in Italian): http://www.commodorefangazette.com

Communities, News, References etc.:

- http://awesome.commodore.me
- http://www.c64.com
- http://c64endings.co.uk

© Springer Science+Business Media Singapore 2015
R. Dillon, *Ready*, DOI 10.1007/978-981-287-341-5

- http://www.cbm8bit.com
- http://codebase64.org
- http://www.commodore.ca
- http://www.commodore64c.com
- http://csdb.dk/
- http://www.lemon64.com

... And many more!

Appendix C
VIC-II Registers

See Table C.1

Table C.1 VIC-II registers

Register	Hex	Decimal	Description
0	D000	53248	X Coordinate Sprite 0
1	D001	53249	Y Coordinate Sprite 0
2	D002	53250	X Coordinate Sprite 1
3	D003	53251	Y Coordinate Sprite 1
4	D004	53252	X Coordinate Sprite 2
5	D005	53253	Y Coordinate Sprite 2
6	D006	53254	X Coordinate Sprite 3
7	D007	53255	Y Coordinate Sprite 3
8	D008	53256	X Coordinate Sprite 4
9	D009	53257	Y Coordinate Sprite 4
10	D00A	53258	X Coordinate Sprite 5
11	D00B	53259	Y Coordinate Sprite 5
12	D00C	53260	X Coordinate Sprite 6
13	D00D	53261	Y Coordinate Sprite 6
14	D00E	53262	X Coordinate Sprite 7
15	D00F	53263	Y Coordinate Sprite 7
16	D010	53264	MSBs of X coordinates
17	D011	53265	Control register 1
18	D012	53266	Raster counter
19	D013	53267	Light Pen X
20	D014	53268	Light Pen Y
21	D015	53269	Sprite enabled

(continued)

© Springer Science+Business Media Singapore 2015
R. Dillon, *Ready*, DOI 10.1007/978-981-287-341-5

Table C.1 (continued)

Register	Hex	Decimal	Description
22	D016	53270	Control register 2
23	D017	53271	Sprite Y expansion
24	D018	53272	Memory pointers
25	D019	53273	Interrupt register
26	D01A	53274	Interrupt enabled
27	D01B	53275	Sprite data priority
28	D01C	53276	Sprite multicolor
29	D01D	53277	Sprite X expansion
30	D01E	53278	Sprite-sprite collision
31	D01F	53279	Sprite-data collision
32	D020	53280	Border color
33	D021	53281	Background color 0
34	D022	53282	Background color 1
35	D023	53283	Background color 2
36	D024	53284	Background color 3
37	D025	53285	Sprite multicolor 0
38	D026	53286	Sprite multicolor 1
39	D027	53287	Color sprite 0
40	D028	53288	Color sprite 1
41	D029	53289	Color sprite 2
42	D02A	53290	Color sprite 3
43	D02B	53291	Color sprite 4
44	D02C	53292	Color sprite 5
45	D02D	53293	Color sprite 6
46	D02E	53294	Color sprite 7

Appendix D
SID Registers

See Table D.1

Table D.1 SID registers

Register	Hex	Decimal	Description
0	D400	54272	Voice #1 frequency (low byte)
1	D401	54273	Voice #1 frequency (hi byte)
2	D402	54274	Voice #1 pulse width (low byte)
3	D403	54275	Voice #1 pulse width (hi byte)
4	D404	54276	**Voice #1 control register**
			Bit #0: 0 = Voice off, Release cycle; 1 = Voice on, Attack-Decay-Sustain cycle
			Bit #1: 1 = Synchronization enabled
			Bit #2: 1 = Ring modulation enabled
			Bit #3: 1 = Disable voice, reset noise generator
			Bit #4: 1 = Triangle waveform enabled
			Bit #5: 1 = Saw waveform enabled
			Bit #6: 1 = Rectangle waveform enabled
			Bit #7: 1 = Noise enabled
5	D405	54277	**Voice #1 Attack and Decay length**
			Bits #0–#3: Decay length. Values (binary, decimal: time)
			0000, 0: 6 ms, 0001, 1: 24 ms, 0010, 2: 48 ms, 0011, 3: 72 ms
			0100, 4: 114 ms, 0101, 5: 168 ms, 0110, 6: 204 ms, 0111, 7: 240 ms
			1000, 8: 300 ms, 1001, 9: 750 ms, 1010, 10: 1.5 s, 1011, 11: 2.4 s
			1100, 12: 3 s, 1101, 13: 9 s, 1110, 14: 15 s, 1111, 15: 24 s
			Bits #4–#7: Attack length. Values: (binary, decimal: time)
			0000, 0: 2 ms, 0001, 1: 8 ms, 0010, 2: 16 ms, 0011, 3: 24 ms
			0100, 4: 38 ms, 0101, 5: 56 ms, 0110, 6: 68 ms, 0111, 7: 80 ms
			1000, 8: 100 ms, 1001, 9: 250 ms, 1010, 10: 500 ms, 1011, 11: 800 ms
			1100, 12: 1 s, 1101, 13: 3 s, 1110, 14: 5 s, 1111, 15: 8 s

(continued)

© Springer Science+Business Media Singapore 2015
R. Dillon, *Ready*, DOI 10.1007/978-981-287-341-5

Table D.1 (continued)

Register	Hex	Decimal	Description
6	D406	54278	**Voice #1 Sustain volume and Release length**
			Bits #0–#3: Release length. Values: (binary, decimal: time)
			0000, 0: 6 ms, 0001, 1: 24 ms, 0010, 2: 48 ms, 0011, 3: 72 ms
			0100, 4: 114 ms, 0101, 5: 168 ms, 0110, 6: 204 ms, 0111, 7: 240 ms
			1000, 8: 300 ms, 1001, 9: 750 ms, 1010, 10: 1.5 s, 1011, 11: 2.4 s
			1100, 12: 3 s, 1101, 13: 9 s, 1110, 14: 15 s, 1111, 15: 24 s
			Bits #4–#7: Sustain volume
7	D407	54279	Voice #2 frequency (low byte)
8	D408	54280	Voice #2 frequency (hi byte)
9	D409	54281	Voice #2 pulse width (low byte)
10	D40A	54282	Voice #2 pulse width (hi byte)
11	D40B	54283	**Voice #2 control register**
			See D404 for meaning of individual bits
12	D40C	54284	**Voice #2 Attack and Decay length**
			See D405 for meaning of individual bits
13	D40D	54285	**Voice #2 Sustain volume and Release length**
			See D406 for meaning of individual bits
14	D40E	54286	Voice #3 frequency (low byte)
15	D40F	54287	Voice #3 frequency (hi byte)
16	D410	54288	Voice #3 pulse width (low byte)
17	D411	54289	Voice #3 pulse width (hi byte)
18	D412	54290	**Voice #3 control register**
			See D404 for meaning of individual bits
19	D413	54291	**Voice #3 Attack and Decay length**
			See D405 for meaning of individual bits
20	D414	54292	**Voice #3 Sustain volume and Release length**
			See D406 for meaning of individual bits
21	D415	54293	Filter cut off frequency (bits #0–#2)
22	D416	54294	Filter cut off frequency (bits #3–#10)
23	D417	54295	Filter control
			Bit #0: 1 = Voice #1 filtered
			Bit #1: 1 = Voice #2 filtered
			Bit #2: 1 = Voice #3 filtered
			Bit #3: 1 = External voice filtered
			Bits #4–#7: Filter resonance
24	D418	54296	Volume and filter modes
			Bits #0–#3: Volume
			Bit #4: 1 = Low pass filter enabled
			Bit #5: 1 = Band pass filter enabled
			Bit #6: 1 = High pass filter enabled
			Bit #7: 1 = Voice #3 disabled

(continued)

Table D.1 (continued)

Register	Hex	Decimal	Description
25	D419	54297	X value of paddle selected at memory address $DD00. (Updates at every 512 system cycles. Read only)
26	D41A	54298	Y value of paddle selected at memory address $DD00. (Updates at every 512 system cycles. Read only)
27	D41B	54299	Voice #3 waveform output. Read only
28	D41C	54300	Voice # 3 ADSR output. Read only

Bibliography

Bagnall, Brian: "Commodore, a company on the edge", Variant Press (2nd Edition), 2010. A very detailed account of everything Commodore, starting from its origins and the MOS acquisition to the rise of Commodore as a leader in the home computer market of the early 1980s.

Dillon, Roberto: "The Golden age of Video Games", CRC Press, 2011. An overview and analysis of the gaming industry and the most groundbreaking video and computer games from the very beginning till the PlayStation launch. Read about Commodore, including the C64 and Amiga, within the context of the whole industry.

Dyer, Sam: "Commodore 64: a Visual Commpendium", Bitmap Books, 2014. A beautiful coffee table book showcasing all the graphical beauty of the most iconic C64 games.

Fisher, Andrew: "The Commodore 64 Book—1982 to 199x", Hiive Books, 2008. A small and delightful book covering over 200 C64 games.

O'Hara, Bob: "Commodork: Sordid tales from a BBS junkie", Lulu.com, 2011. Humorous and engaging recollections of living online before the internet.

Tomczyk, Michael: "Home Computer Wars", Compute Books, 1984. A unique account of the life at Commodore in the early 1980s and the launch of the VIC-20, directly from the trenches.

© Springer Science+Business Media Singapore 2015
R. Dillon, *Ready*, DOI 10.1007/978-981-287-341-5

Index

© Springer Science+Business Media Singapore 2015
R. Dillon, *Ready*, DOI 10.1007/978-981-287-341-5

Printed in the United States
By Bookmasters